McGraw-Hill

Mathematics

Daily Homework Practice

2

McGraw-Hill School Division
New York Farmington

McGraw-Hill School Division
A Division of The ***McGraw-Hill*** *Companies*

McGraw-Hill School Division
Two Penn Plaza
New York, New York 10121-2298

Printed in the United States of America

ISBN 0-02-100284-3 / 2

4 5 6 7 8 9 024 05 04 03 02 01

GRADE 2
Contents

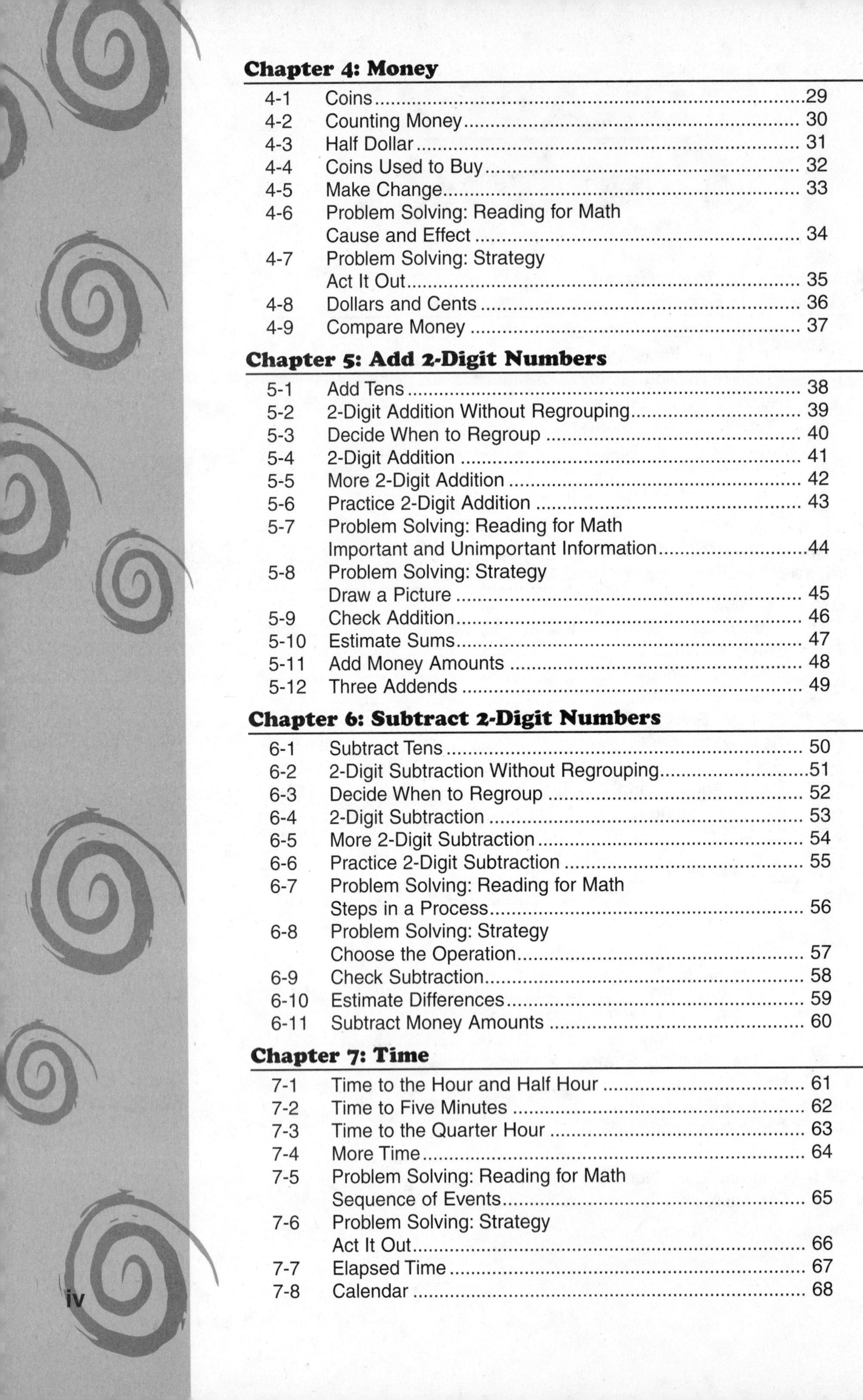

Chapter 4: Money

Chapter 5: Add 2-Digit Numbers

Chapter 6: Subtract 2-Digit Numbers

Chapter 7: Time

Chapter 8: Data and Graphs

Chapter 9: Measurement

Chapter 10: Geometry

Chapter 11: Fractions and Probability

Chapter 12: Place Value to 1,000

Chapter 13: Add and Subtract 3-Digit Numbers

Chapter 14: Multiplication and Division

Name ______________________________

Addition Strategies

Add.

Start with the greater number.

1. 5 + 3 = 8   4 + 0 = ____   7 + 2 = ____

2. 6 + 1 = ____   3 + 2 = ____   9 + 3 = ____

3.

8	9	10	6	5	6	8
+ 3	+ 0	+ 2	+ 1	+ 2	+ 3	+ 0

Problem Solving

4. There are 6 snails in the pond. 3 more snails slip into the pond. How many snails are in the pond now?

____ snails

5. There are 4 ducks in the pond. 2 more ducks jump into the pond. How many ducks are there in all?

____ ducks

Spiral Review

Add.

6.

2	3	5	4	3	2	4
+ 2	+ 2	+ 1	+ 3	+ 1	+ 0	+ 2

Name ____________________

Turnaround Facts

Find each sum.

1.	4 + 3 = 7	3 + 4 = 7	4 + 6	6 + 4	2 + 5	5 + 2
2.	3 + 8	8 + 3	4 + 5	5 + 4	3 + 9	9 + 3
3.	3 + 7	7 + 3	8 + 4	4 + 8	6 + 5	5 + 6

Are these turnaround facts?
Write *yes* or *no*.

4.	4 + 5 5 + 4 ____	4 + 6 6 + 6 ____	7 + 4 4 + 7 ____

Spiral Review

Find each sum.

5. 7 + 3 ____ 8 + 2 ____ 4 + 1 ____ 10 + 2 ____ 2 + 8 ____ 6 + 3 ____ 5 + 1 ____

Name ______________________________

1-3 Problem Solving: Reading for Math

Read to Set a Purpose

Kelly saw 5 frogs in the pond.
Matt saw 3 frogs in the grass.

Solve.

1. How many frogs did Kelly see in the pond? ______ frogs

2. How many frogs did Matt see in the grass? ______ frogs

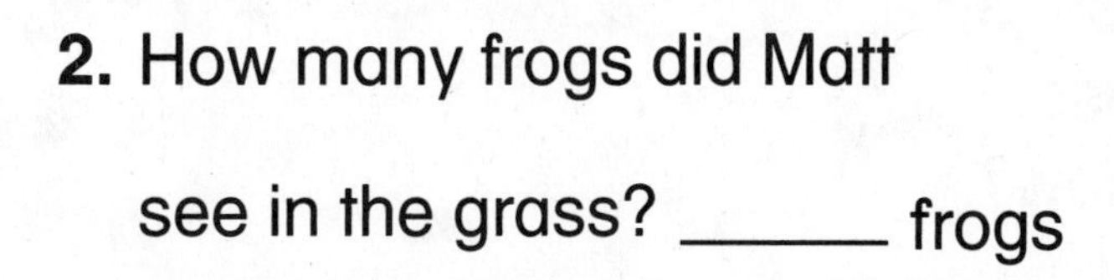

3. How many frogs did they see in all? ______ frogs

   Write a number sentence to show your thinking.

   ____

4. What information helped you find the answer?

   ____

Spiral Review

Add.

5.

8	5	10	7	6	9	4
+ 2	+ 3	+ 2	+ 3	+ 1	+ 3	+ 2

Name ______________________________

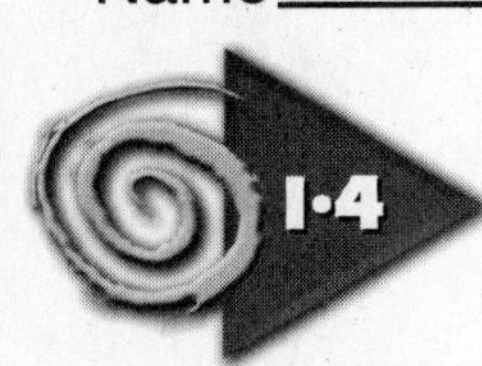

Problem Solving: Strategy
Draw a Picture

Draw a picture. Solve.

1. There are 2 ducks on the pond. 4 more ducks land on the pond. How many ducks are there now?

   ________ ducks

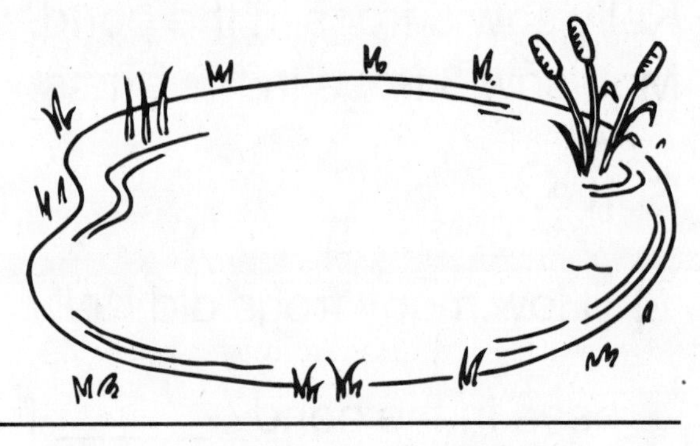

2. There are 9 ants in the grass. 3 more ants walk into the grass. How many ants are there now?

   ________ ants

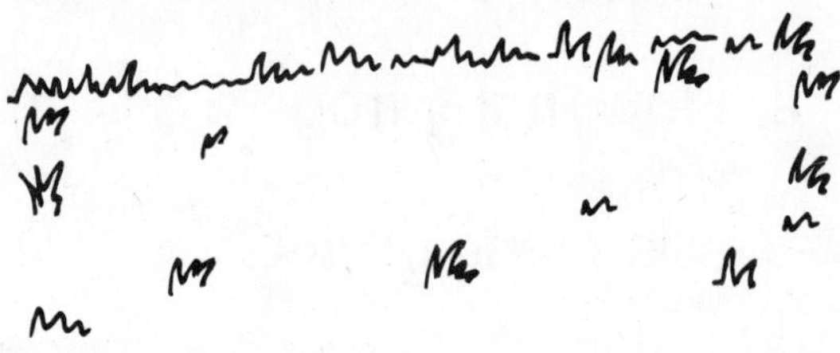

3. There are 5 bugs on a log. 4 more bugs fly onto the log. How many bugs are there now?

   ________ bugs

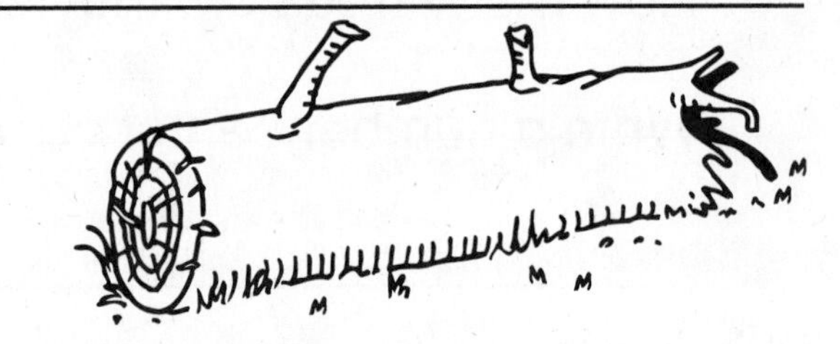

4. There are 5 rabbits in a field. 3 more rabbits hop into the field. How many rabbits are there now?

   ________ rabbits

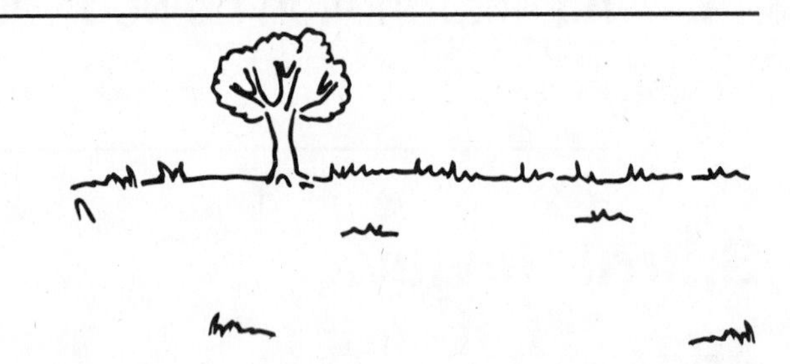

Spiral Review

Add.

5. $\begin{array}{r} 8 \\ +3 \\ \hline \end{array}$ $\begin{array}{r} 8 \\ +0 \\ \hline \end{array}$ $\begin{array}{r} 7 \\ +3 \\ \hline \end{array}$ $\begin{array}{r} 6 \\ +1 \\ \hline \end{array}$ $\begin{array}{r} 3 \\ +9 \\ \hline \end{array}$ $\begin{array}{r} 3 \\ +7 \\ \hline \end{array}$ $\begin{array}{r} 5 \\ +3 \\ \hline \end{array}$

Name ____________________

Count Back to Subtract

Subtract. You can use a number line to count back.

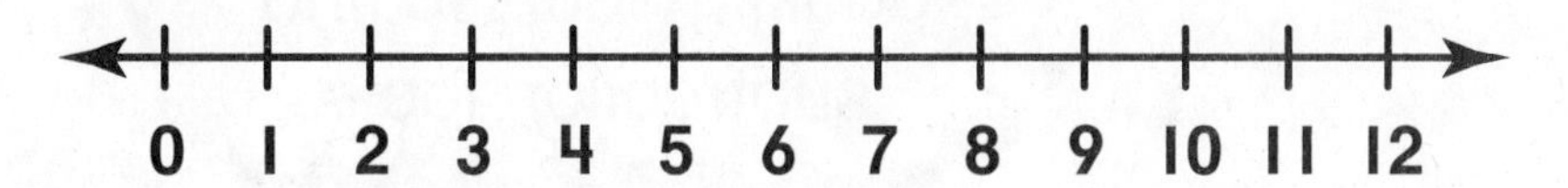

1. 9 − 3 = 6 　　12 − 1 = ____ 　　7 − 2 = ____

2. 5 − 3 = ____ 　　8 − 2 = ____ 　　12 − 3 = ____

3. 9 − 2 = ____ 　　11 − 3 = ____ 　　10 − 3 = ____

4.

9	7	6	0	11	6
− 1	− 3	− 1	− 0	− 2	− 2

5.

5	12	5	9	10	5
− 0	− 2	− 2	− 0	− 2	− 2

Problem Solving

6. 8 frogs sit on a log. 3 frogs hop away. How many frogs are left?

____ frogs

8

Spiral Review

Add.

7.

5	9	6	8	3	7	10
+ 2	+ 3	+ 1	+ 0	+ 8	+ 2	+ 1

Name________________________________

Relate Addition and Subtraction

Add. Write a related subtraction sentence.

Remember you can use addition facts to find subtraction facts.

1. 7 + 5 = 12   3 + 6 = ____   6 + 4 = ____
   12 − 7 = 5   ___ − ___ = ___   ___ − ___ = ___

2. 5 + 3 = ____   5 + 6 = ____   3 + 4 = ____
   ___ − ___ = ___   ___ − ___ = ___   ___ − ___ = ___

3. 9 + 3 = ____   5 + 4 = ____   4 + 7 = ____
   ___ − ___ = ___   ___ − ___ = ___   ___ − ___ = ___

Is the addition related to the subtraction? Write *yes* or *no*.

4. 3 + 6 = 9   3 + 5 = 8   5 + 7 = 12
   9 − 6 = 3   5 − 2 = 3   12 − 5 = 7
   ______   ______   ______

Spiral Review

Add or subtract.

5. $\begin{array}{r} 8 \\ +2 \\ \hline \end{array}$ $\begin{array}{r} 12 \\ -8 \\ \hline \end{array}$ $\begin{array}{r} 6 \\ +3 \\ \hline \end{array}$ $\begin{array}{r} 10 \\ -3 \\ \hline \end{array}$ $\begin{array}{r} 8 \\ -0 \\ \hline \end{array}$ $\begin{array}{r} 5 \\ +1 \\ \hline \end{array}$ $\begin{array}{r} 7 \\ +3 \\ \hline \end{array}$

Name ________________________________

1·7 Fact Families

Add or subtract.
Complete each fact family.

1.

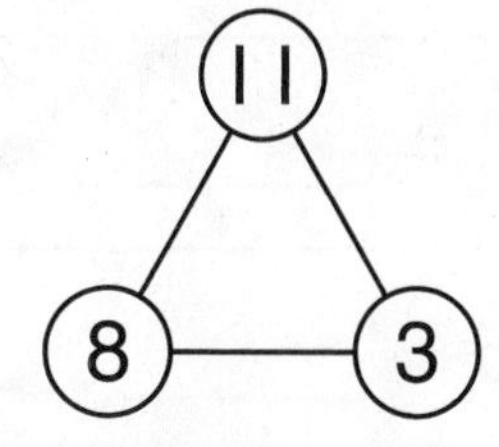

8 + 3 = ______ 11 − 8 = ______

3 + ______ = 11 ______ − 3 = 8

2.

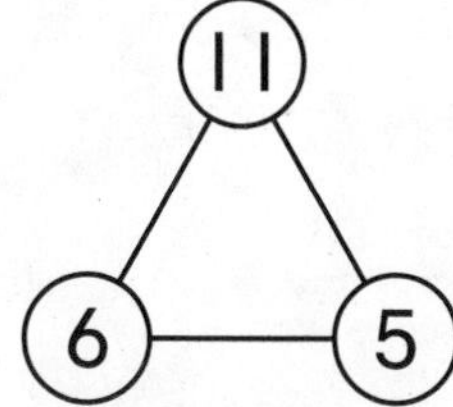

6 + 5 = ______ 11 − ______ = 6

5 + ______ = 11 11 − 6 = ______

Complete each fact family triangle.

3.

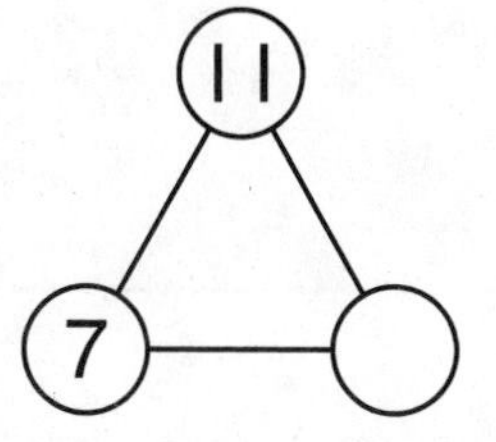

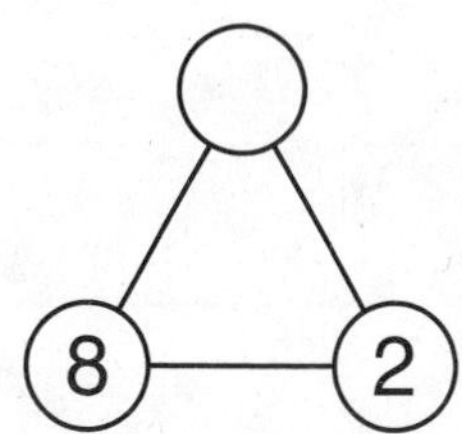

8 2

4.

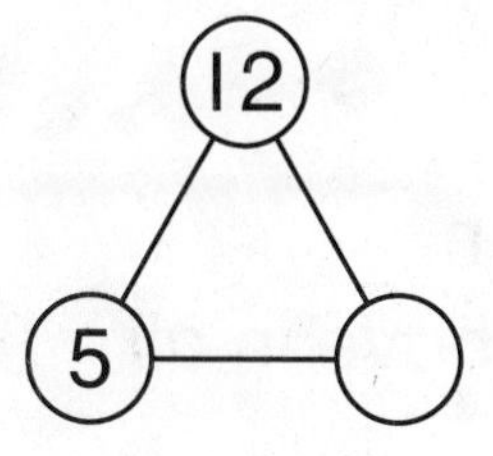

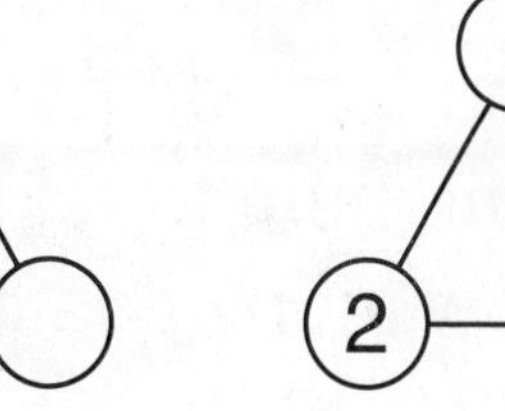

2 9

Spiral Review

Complete each number sentence.

5. 9 + 2 = ______ 12 − ______ = 7 10 − 7 = ______

6. 6 + ______ = 10 10 − 0 = ______ ______ − 1 = 7

Name

Doubles to Add

Add.

1. 5 + 5 = 10 5 + 6 = ___ 5 + 4 = ___

2. 3 + 3 = ___ 3 + 2 = ___ 3 + 4 = ___

3. 4 + 4 = ___ 4 + 3 = ___ 4 + 5 = ___

4. $\begin{array}{r} 6 \\ +6 \\ \hline \end{array}$ $\begin{array}{r} 6 \\ +7 \\ \hline \end{array}$ $\begin{array}{r} 6 \\ +5 \\ \hline \end{array}$ $\begin{array}{r} 7 \\ +7 \\ \hline \end{array}$ $\begin{array}{r} 7 \\ +8 \\ \hline \end{array}$ $\begin{array}{r} 7 \\ +6 \\ \hline \end{array}$

5. $\begin{array}{r} 8 \\ +8 \\ \hline \end{array}$ $\begin{array}{r} 8 \\ +9 \\ \hline \end{array}$ $\begin{array}{r} 8 \\ +7 \\ \hline \end{array}$ $\begin{array}{r} 9 \\ +9 \\ \hline \end{array}$ $\begin{array}{r} 9 \\ +10 \\ \hline \end{array}$ $\begin{array}{r} 9 \\ +8 \\ \hline \end{array}$

Problem Solving

6. Pete has 6 Grand Canyon postcards. Lisa has the same number of postcards from Yellowstone. How many postcards do they have in all?

___

Spiral Review

Complete each number sentence.

7. 6 + 5 = ___ ___ − 6 = 5 6 + ___ = 11

Name ___________________________________

Make a Ten to Add 7, 8, 9

Add. You can use  **and the 10-frame.**

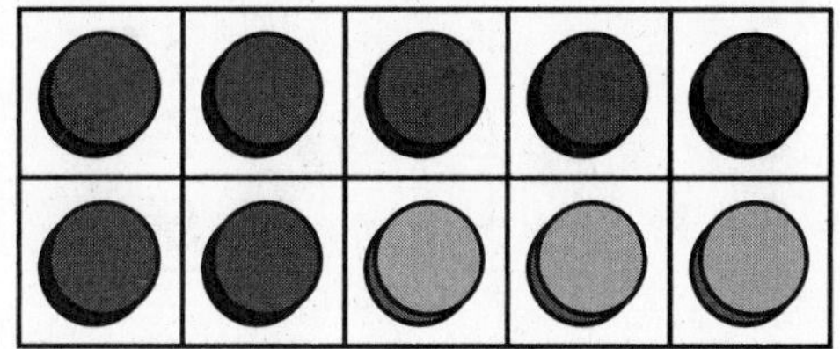
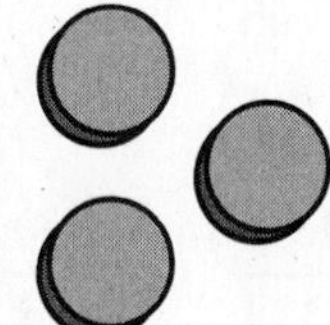

1. 7 + 6 = 13

2. 9 + 4 = ______   6 + 8 = ______   3 + 7 = ______

3. 8 + 5 = ______   2 + 9 = ______   7 + 7 = ______

4.
$$\begin{array}{r} 8 \\ +3 \\ \hline \end{array} \quad \begin{array}{r} 6 \\ +9 \\ \hline \end{array} \quad \begin{array}{r} 5¢ \\ +7¢ \\ \hline \end{array} \quad \begin{array}{r} 2 \\ +8 \\ \hline \end{array} \quad \begin{array}{r} 9 \\ +5 \\ \hline \end{array} \quad \begin{array}{r} 7 \\ +6 \\ \hline \end{array} \quad \begin{array}{r} 4¢ \\ +8¢ \\ \hline \end{array}$$

5.
$$\begin{array}{r} 4 \\ +7 \\ \hline \end{array} \quad \begin{array}{r} 9¢ \\ +3¢ \\ \hline \end{array} \quad \begin{array}{r} 5 \\ +9 \\ \hline \end{array} \quad \begin{array}{r} 6 \\ +7 \\ \hline \end{array} \quad \begin{array}{r} 7 \\ +3 \\ \hline \end{array} \quad \begin{array}{r} 5 \\ +8 \\ \hline \end{array} \quad \begin{array}{r} 8 \\ +6 \\ \hline \end{array}$$

Solve.

6. Matt saw 5 kittens and 7 puppies.
How many animals did he see?

_________ animals

Spiral Review

Add.

7.
$$\begin{array}{r} 8 \\ +8 \\ \hline \end{array} \quad \begin{array}{r} 8 \\ +7 \\ \hline \end{array} \quad \begin{array}{r} 6 \\ +6 \\ \hline \end{array} \quad \begin{array}{r} 7 \\ +6 \\ \hline \end{array} \quad \begin{array}{r} 7 \\ +7 \\ \hline \end{array} \quad \begin{array}{r} 7 \\ +8 \\ \hline \end{array} \quad \begin{array}{r} 9 \\ +9 \\ \hline \end{array}$$

Name____

Three Addends

Add. Use different strategies.

1.	6	5	6	3	3	2	6
	6	8	2	7	4	7	3
	+ 2	+ 5	+ 4	+ 5	+ 4	+ 8	+ 3
	14						

2.	7	3	8	4	5	9	8
	2	9	8	6	5	8	2
	+ 7	+ 1	+ 3	+ 5	+ 6	+ 1	+ 3

3.	1	4	5	6	8	6	3
	7	9	7	5	1	6	5
	+ 3	+ 4	+ 7	+ 6	+ 8	+ 4	+ 7

4. 5 + 5 + 2 = ______ 9 + 6 + 4 = ______ 3 + 8 + 7 = ______

Solve.

5. Karen bought 3 postcards on the first day of her trip. She bought 6 postcards on the second day, and 5 postcards on the third day. How many postcards did she buy in all?

__________ postcards

Spiral Review

Add or subtract.

6.	4	11	8	12	12	9	14
	+ 7	− 4	+ 4	− 8	− 4	+ 5	− 5

Name ___________________________________

Problem Solving: Reading for Math
Use a Summary

Dear Grandpa,
The Grand Canyon is neat!
I took a walk to learn about the plants and animals here.
There were 5 grown-ups on the walk.
There were 8 kids.
Wish you were here.

Love, Pete

Mr. Joseph Reyes
100 Smith Ave.
Mapleville, MA
12345

Solve.

1. Write 2 important number facts from the story.

____

2. Use both facts to write one sentence that tells a number story.

____

3. How many people went on the walk in all? ____________

Write a number sentence to show your thinking.

____

Spiral Review

Add.

4.

8	8	1	2	6	8	6
2	2	3	7	6	9	9
+ 8	+ 7	+ 7	+ 7	+ 3	+ 1	+ 4

Name ______________________________

Problem Solving: Strategy
Write a Number Sentence

Write a number sentence. Solve.

1. Kate has 9 postcards. Toby has 5 postcards. How many more postcards does Kate have than Toby?

_____ ◯ _____ = _____ postcards

2. Jenny buys 5 fish postcards and 8 bear postcards. How many postcards does she have altogether?

_____ ◯ _____ = _____ postcards

3. Mike has 12 postcards. He mails 8 postcards. How many does he have left?

_____ ◯ _____ = _____ postcards

4. Carla buys 2 stamps. Each stamp costs 6¢. How much does Carla spend?

_____ ◯ _____ = _____

Spiral Review

Add or subtract.

5. $\begin{array}{r} 5 \\ +\ 4 \\ \hline \end{array}$ $\begin{array}{r} 9 \\ -\ 6 \\ \hline \end{array}$ $\begin{array}{r} 2 \\ +\ 6 \\ \hline \end{array}$ $\begin{array}{r} 10 \\ -\ 5 \\ \hline \end{array}$ $\begin{array}{r} 7 \\ +\ 3 \\ \hline \end{array}$ $\begin{array}{r} 10 \\ -\ 2 \\ \hline \end{array}$ $\begin{array}{r} 3 \\ +\ 4 \\ \hline \end{array}$

Name ______________________________

Doubles to Subtract

Add or subtract.

1. $8 + 8 =$ 16 $\qquad 16 - 8 =$ ______

2. $10 + 10 =$ ______ $\qquad 20 - 10 =$ ______

3. $5 + 5 =$ ______ $\qquad 10 - 5 =$ ______

4. $6 + 6 =$ ______ $\qquad 12 - 6 =$ ______

5. $\begin{array}{r} 7 \\ +\ 7 \\ \hline \end{array}$ $\begin{array}{r} 14 \\ -\ 7 \\ \hline \end{array}$ | $\begin{array}{r} 9 \\ +\ 9 \\ \hline \end{array}$ $\begin{array}{r} 18 \\ -\ 9 \\ \hline \end{array}$ | $\begin{array}{r} 10 \\ +\ 10 \\ \hline \end{array}$ $\begin{array}{r} 20 \\ -\ 10 \\ \hline \end{array}$

6. $\begin{array}{r} 6 \\ +\ 6 \\ \hline \end{array}$ $\begin{array}{r} 12 \\ -\ 6 \\ \hline \end{array}$ | $\begin{array}{r} 8 \\ +\ 8 \\ \hline \end{array}$ $\begin{array}{r} 16 \\ -\ 8 \\ \hline \end{array}$ | $\begin{array}{r} 5 \\ +\ 5 \\ \hline \end{array}$ $\begin{array}{r} 10 \\ -\ 5 \\ \hline \end{array}$

Solve.

7. It is 5 miles from Ashley's house to the mall. How many miles is the drive from Ashley's house to the mall and back home again?

   ________ miles

Spiral Review

Add.

8. $\begin{array}{r} 7 \\ +\ 4 \\ \hline \end{array}$ $\begin{array}{r} 4 \\ +\ 7 \\ \hline \end{array}$ $\begin{array}{r} 8 \\ +\ 2 \\ \hline \end{array}$ $\begin{array}{r} 2 \\ +\ 8 \\ \hline \end{array}$ $\begin{array}{r} 9 \\ +\ 3 \\ \hline \end{array}$ $\begin{array}{r} 3 \\ +\ 9 \\ \hline \end{array}$

Name ______________________________

2·7 Relate Addition and Subtraction

Complete each number sentence.

1. 8 + 7 = 15   9 + 9 = ____   7 + 6 = ____

15 − 8 = 7   18 − 9 = ____   13 − 7 = ____

2. 7 + 6 = ____   9 + 8 = ____   7 + 7 = ____

13 − 7 = ____   17 − 9 = ____   14 − 7 = ____

3. 8 + 5 = ____   13 − 8 = ____   9 + 5 = ____   14 − 9 = ____   9 + 6 = ____   15 − 9 = ____

Problem Solving

Use the graph.

4. How many stamps are there in all?

____ stamps

5. How many more flower stamps are there than fish stamps?

____ stamps

Spiral Review

Add.

6. 3 + 7 + 4 = ____   2 + 7 + 7 = ____   5 + 3 + 5 = ____   2 + 7 + 8 = ____   6 + 1 + 6 = ____   6 + 4 + 2 = ____   1 + 5 + 9 = ____

Name ______________________________

Missing Addends

Find each missing number.

1. 9 + [7] = 16 16 − 9 = [ ]

2. 8 + [ ] = 14 14 − 8 = [ ]

3.

8	15	7	13	9	13
+ [ ]	− 8	+ [ ]	− 7	+ [ ]	− 9
15		13		13	

Complete each number sentence.

4. 9 + [ ] = 15 8 + [ ] = 13 9 + [ ] = 14

5. 7 ◯ 8 = 15 13 ◯ 5 = 8 7 ◯ 9 = 16

Find the doubles.

6. [ ] + [ ] = 8 [ ] + [ ] = 4

7. [ ] + [ ] = 6 [ ] + [ ] = 2

Spiral Review

Add.

8. 7 + 4 = ______ 8 + 5 = ______ 9 + 3 = ______

9. 8 + 4 = ______ 8 + 6 = ______ 9 + 4 = ______

Name ____________________

2·9 Fact Families

Complete each fact family.

1.

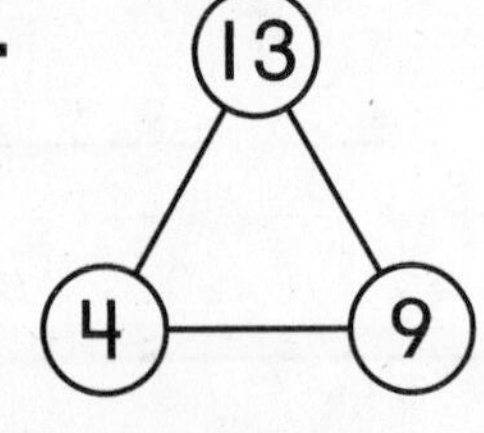

9 + 4 = 13 4 + 9 = ____

13 − 9 = ____ 13 − 4 = ____

2.

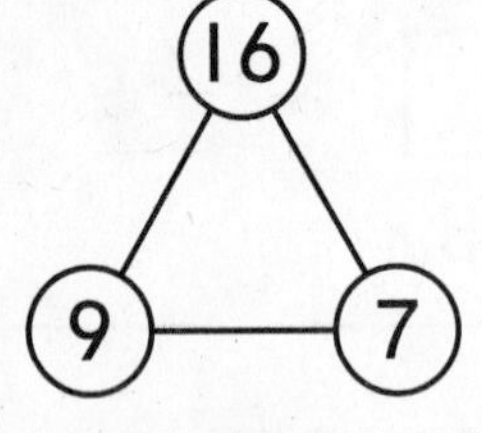

9 + 7 = ____ 7 + 9 = ____

16 − 9 = ____ 16 − 7 = ____

3.

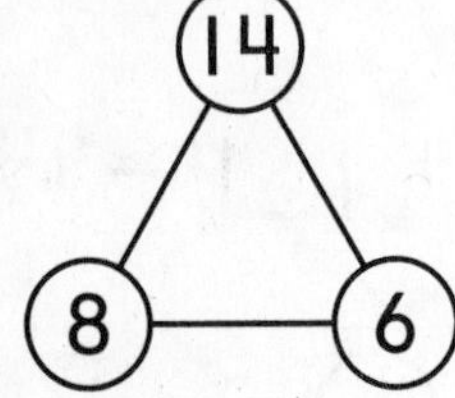

8 + 6 = ____ 6 + ____ = 14

14 ◯ 8 = 6 14 − ____ = 8

4.

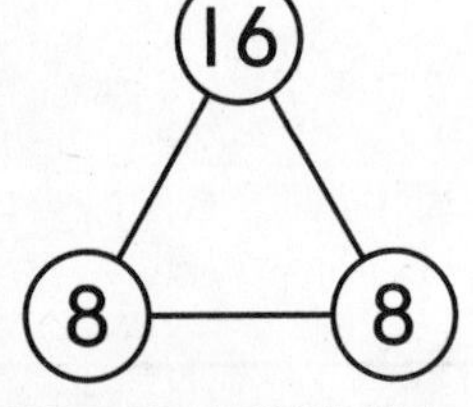

8 ◯ 8 = 16 16 ◯ 8 = 8

Spiral Review

Draw a picture to solve this problem.

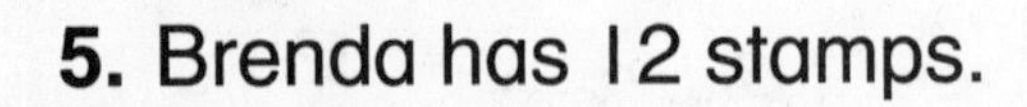

5. Brenda has 12 stamps.

She gives 4 stamps to Josh.

How many stamps does Brenda have left? ____ stamps

Name ____________________

Tens

Write how many.

1. 2 groups of ten

2 tens =

20 in all

4 groups of ten

_____ tens =

_____ in all

2. 6 groups of ten

_____ tens =

_____ in all

5 groups of ten

_____ tens =

_____ in all

3. 7 groups of ten

_____ tens =

_____ in all

9 groups of ten

_____ tens =

_____ in all

4. 10 groups of ten

_____ tens =

_____ in all

8 groups of ten

_____ tens =

_____ in all

5. 3 groups of ten

_____ tens =

_____ in all

1 group of ten

_____ ten =

_____ in all

Spiral Review

Subtract.

6. 11 − 8 = _____ 10 − 6 = _____ 12 − 7 = _____

Name ____________________

More Tens

About how many are in each group? Circle your estimate.

1. 

about 10 (about 20)

2. 

about 30 about 60

3. 

about 10 about 40

4. 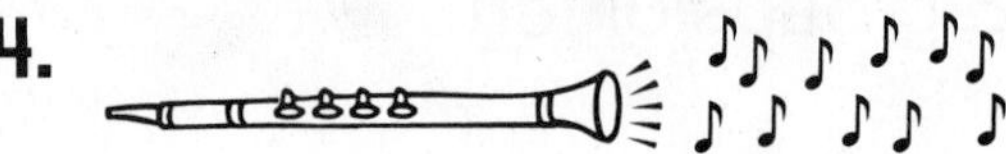

about 10 about 30

Problem Solving

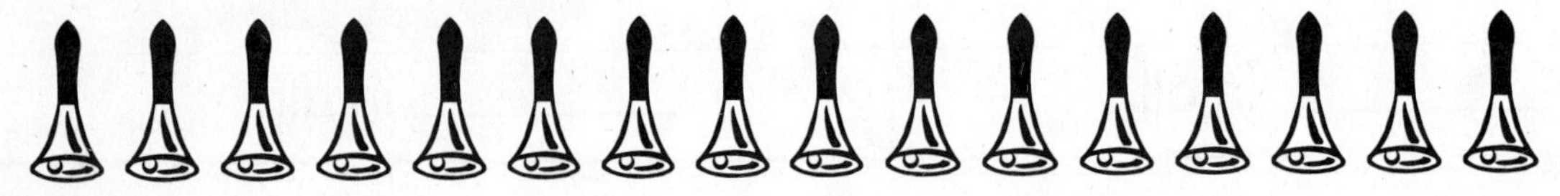

5. About how many bells are in the picture? ____________

6. Now count how many. ________

Spiral Review

Solve. Write a number sentence for each problem.

7. There are 6  in the class. There are 3 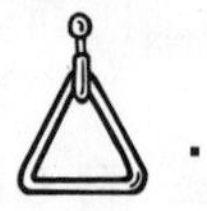.

How many  and  are there all together?

____

________ ◯ ________ = ________

Name______________________

Tens and Ones

Write how many tens and ones.

1. 36 = __3__ tens __6__ ones

tens	ones
3	6

2. 72 = ______ tens ______ ones

tens	ones

3. 45 = ______ tens ______ ones

tens	ones

Write each number.

4. 9 tens 4 ones ______ 1 ten and 8 ones ______

Problem Solving

Draw a picture to solve.

5. Liz has 4 tens and 2 ones Mark has 2 tens and 2 ones. How many more tens does Liz have than Mark?

______

Spiral Review

Add or subtract.

6. 5 + 5 = ______ 16 − 8 = ______ 9 + 9 = ______

Name ____________________

Numbers to 100

Write each number.

1. six 6	eighteen ______	forty ______
2. seventy ______	sixty-four ______	eighty-two ______
3. twenty-five ______	fifty-one ______	ninety ______
4. thirty-six ______	ninety-three ______	thirty-nine ______

Write each number word.

5. 8 ______________	13 ______________
6. 60 ______________	44 ______________
7. 27 ______________	95 ______________
8. 76 ______________	38 ______________

Spiral Review

Complete each number sentence.

9. 3 + 9 = ______ 7 + 4 = ______ 8 + 2 = ______

10. 3 + ______ = 12 7 + ______ = 11 8 + ______ = 10

Name ________________________________

Expanded Form

Write each number in expanded form.

1. **89** 80 + 9 ______ **24** ______

2. **12** ______ **61** ______

3. **35** ______ **47** ______

4. **76** ______ **98** ______

Spiral Review

Add or subtract.

5.

8	7	18	16	4	9	17
+ 7	+ 6	− 9	− 7	+ 8	+ 9	− 9

Name________________________________

3·6 Problem Solving: Reading for Math

Draw Conclusions

Read. Look at the pictures. Answer each question.

1. Are there more children making the meal or setting the table?

____

2. Does everyone have a job to do? Why?

____

____

Spiral Review

Write each number in expanded form.

3. 83   46   25   91

________   ________   ________   ________

Name ____________________

Problem Solving: Strategy
Use Logical Reasoning

Solve.

1. How many candy bars did Joe sell?
   The number of candy bars is between 20 and 30.
   The ones digit is just after 8.
   ___29___ candy bars

2. How many meals were served at lunch?
   The number has a 7 in the tens place.
   The digit in the ones place is greater than 2.
   The number is less than 74.
   ________ meals

3. How many flowers did the class plant?
   The number is greater than 40.
   The number is less than 50.
   The tens and ones digits are the same.
   ________ flowers

Spiral Review

Write each number.

4. four ______ seventeen ______ eighty-five ______

Write each number word.

5. 7 __________ 15 __________ 32 __________

Name ______________________________

3·8 Compare Numbers

Compare.

Use <, >, or =.

Use ◻ if you like.

Remember to compare the tens first.

1. 58 __>__ 57 45 ____ 45 42 ____ 9 33 ____ 44

2. 3 ____ 34 98 ____ 92 13 ____ 2 65 ____ 25

3. 22 ____ 22 35 ____ 64 19 ____ 27 5 ____ 63

4. 81 ____ 26 17 ____ 71 89 ____ 89 99 ____ 88

5. 49 ____ 38 26 ____ 26 78 ____ 82 18 ____ 19

6. 36 ____ 29 12 ____ 76 44 ____ 46 77 ____ 67

Spiral Review

Complete each number sentence.

7. 9 + 6 = ________

9 + ________ = 15

15 − 9 = ________

15 ◯ 6 = 9

7 + 5 = ________

7 ◯ 5 = 12

12 − 7 = ________

12 − ________ = 7

Name________________________________

Order Numbers

1	2	3	4	5	6	7	8	9	10
11	12	13	14	15	16	17	18	19	20
21	22	23	24	25	26	27	28	29	30
31	32	33	34	35	36	37	38	39	40
41	42	43	44	45	46	47	48	49	50
51	52	53	54	55	56	57	58	59	60
61	62	63	64	65	66	67	68	69	70
71	72	73	74	75	76	77	78	79	80
81	82	83	84	85	86	87	88	89	90
91	92	93	94	95	96	97	98	99	100

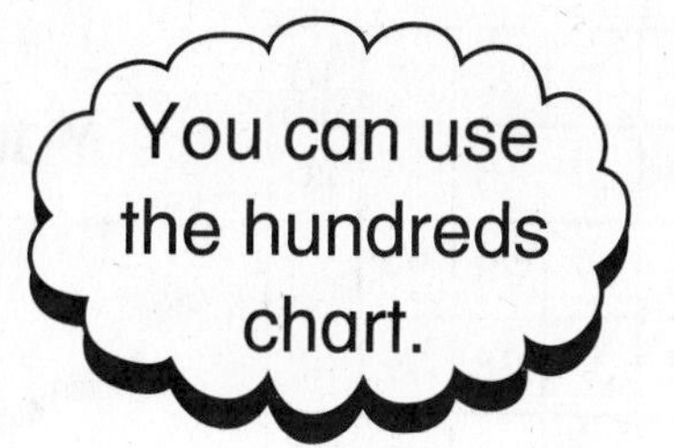

Write the number that comes just before.

1. ________, 66　　________, 40　　________, 75

2. ________, 57　　________, 96　　________, 41

Write the number just after.

3. 74, ________　　27, ________　　91, ________

Write the number that comes between.

4. 34, ________, 36　　60, ________, 62　　13, ________, 15

Spiral Review

Add.

5. $5 + 6$　　$6 + 5$　|　$7 + 5$　　$5 + 7$　|　$6 + 4$　　$4 + 6$

Name______________________________

Skip Count

1	2	3	4	5	6	7	8	9	10
11	12	13	14	15	16	17	18	19	20
21	22	23	24	25	26	27	28	29	30
31	32	33	34	35	36	37	38	39	40
41	42	43	44	45	46	47	48	49	50
51	52	53	54	55	56	57	58	59	60
61	62	63	64	65	66	67	68	69	70
71	72	73	74	75	76	77	78	79	80
81	82	83	84	85	86	87	88	89	90
91	92	93	94	95	96	97	98	99	100

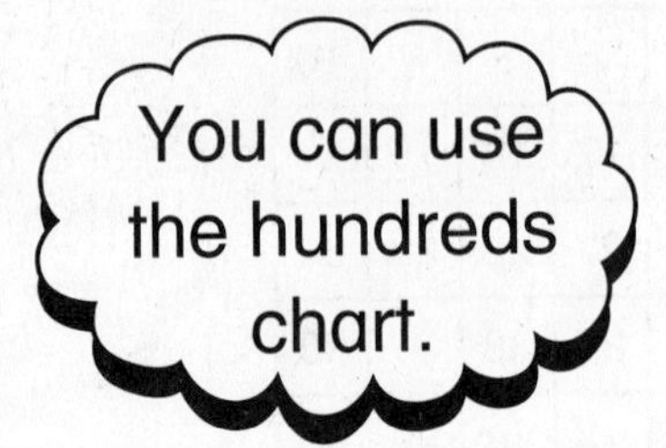

1. Skip count by twos. Write the missing numbers.

2 4 ______ 8 ______ 12 ______ ______

2. Skip count by threes. Write the missing numbers.

3 ______ 9 ______ ______ 18 ______ ______

3. Skip count by fours. Write the missing numbers.

4 8 ______ ______ ______ 24 28 ______

4. Skip count by fives. Write the missing numbers.

5 10 ______ ______ ______ 30 ______ 40

Spiral Review

Compare. Use <, >, and =.

5. 36 ____ 54 29 ____ 29 56 ____ 9 76 ____ 67

Name ______________________________

Odd and Even Numbers

Write if the number is even or odd.

1. 55 ___odd___ 62 ________ 16 ________

2. 43 ________ 21 ________ 94 ________

3. 60 ________ 87 ________ 38 ________

Write the next three even numbers.

4. 20 ______ ______ ______ | 72 ______ ______ ______

5. 46 ______ ______ ______ | 88 ______ ______ ______

Write the next three odd numbers.

6. 65 ______ ______ ______ | 17 ______ ______ ______

7. 33 ______ ______ ______ | 81 ______ ______ ______

Spiral Review

About how many notes are in each group? Circle your estimate.

8.

about 10 about 20

9.

about 30 about 60

Name ____________________

3-12 Ordinal Numbers

Follow each direction.

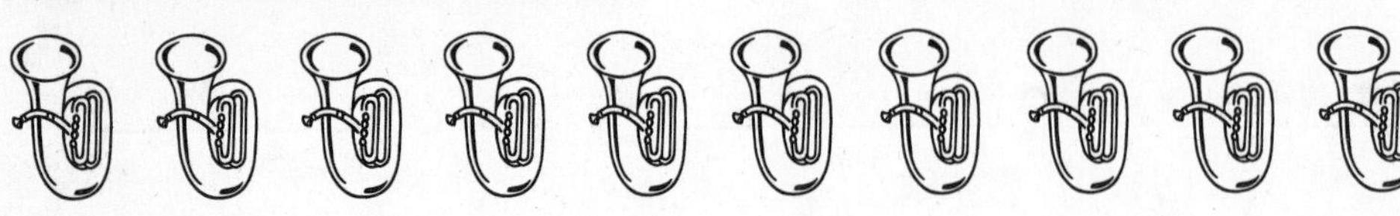

first

1. Color the second  red.

2. Color the eighth  blue.

3. Color the tenth  yellow.

4. Color the sixth  green.

5. How many  are in front of the seventh ?

______

6. How many  are in front of the third ?

______

Answer each question.

7. There are 12 children in the lunch line. How many are in front of the ninth child?

______

8. There are 12 children in the lunch line. How many are behind the fourth child?

______

Spiral Review

Write the number that comes just before.

9. ______, 54 ______, 26 ______, 93

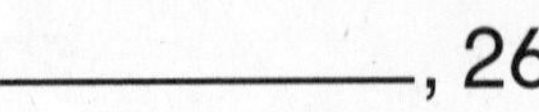

Name ______________________________

Coins

Use coins.

Show each price two ways.

Draw the coins.

penny

nickel

dime

quarter

1. 

2. 

3. 

Problem Solving

4. Kate and Tim bought a 

. The 

cost 17¢. Tim paid 8¢.

Kate paid the rest. How much did Kate pay? __________

Spiral Review

Subtract.

5.

20	17	15	13	16	18	14
− 10	− 8	− 7	− 8	− 9	− 9	− 7

Name ______________________________

Counting Money

Count to find the total amount.

Do you have enough money to buy each item?

Circle yes or no.

no

1. 

25 50 60 65

---

2. 

yes no

_____ _____ _____ _____ _____ _____

---

3. 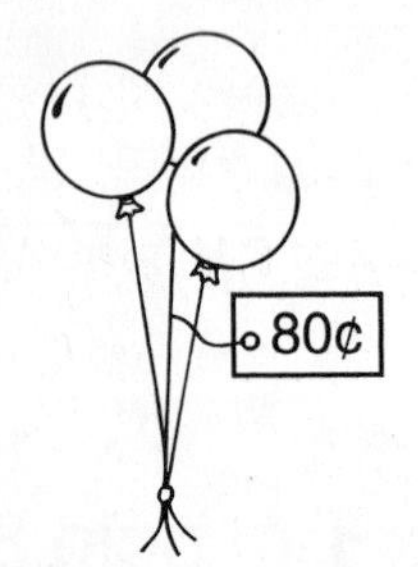

yes no

_____ _____ _____ _____ _____ _____

Spiral Review

Add.

4.

6	2	7	6	2	3	7
8	5	3	4	4	3	5
+ 4	+ 5	+ 5	+ 6	+ 8	+ 7	+ 7

Name____________________________________

Half Dollar

Count each group of coins.

Write the total amount.

1. 

 64¢

2. 

____

3. 

____

4. 

____

Circle the coins for the amount.

5. 

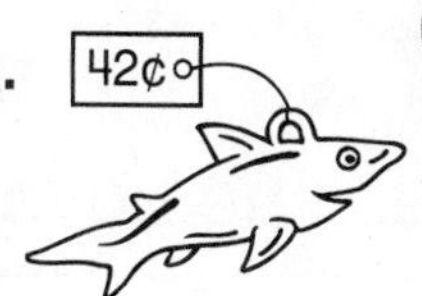

6.. 

Problem Solving

7. Ted has   . Carla has  .

How much do they have altogether? ____ cents

Spiral Review

Write if the number is even or odd.

8. 25 ______ 58 ______ 34 ______ 63 ______

Name ______________________________

Coins Used to Buy

Use the fewest number of coins to pay.

Circle the coins.

1. 

2. 

3. 

Problem Solving

4. Beth has  . She buys 

.

How much does she have left? __________ cents

Spiral Review

Write how many tens and ones.

5. 34 =

tens	ones
3	4

6. 15 =

tens	ones

7. 66 =

tens	ones

8. 79 =

tens	ones

Name ______________________________

Make Change

Count up to find the change.

	Price	You pay	Change
1.			4¢ change
2.			______ change
3.			______ change

Problem Solving

4. Which set of coins has the greater value? Explain.

____

____

Spiral Review

5. Skip count by fives.

5 10 ______ ______ 25 ______ ______ ______

Name________________________________

4•6 Problem Solving: Reading for Math
Cause and Effect

Josh wants to go to the fair.

He helps the lady next door sweep the walk and put out the trash.

She gives Josh 1 quarter, 2 dimes, and 1 nickel.

1. How does Josh help the lady next door?

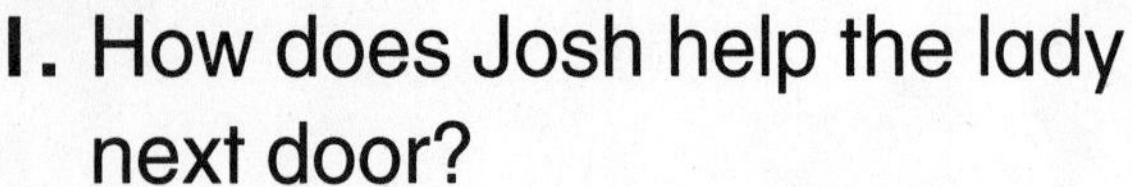

____

2. Why does Josh help the lady next door?

____

3. What does Josh get for helping?

____

4. How much money does Josh get? ________________

Spiral Review

Follow each direction.

5. Color the third duck red.

6. Color the eighth duck yellow.

7. Color the tenth duck blue.

Name ______________________________

Problem Solving: Strategy
Act It Out

Use coins. Act it out.

1. Kate has quarters, dimes, and nickels. Show 3 ways she can buy an ice cream cone for 35¢.

   ______ quarters ______ dimes ______ nickels

   ______ quarters ______ dimes ______ nickels

   ______ quarters ______ dimes ______ nickels

2. Beth has 6 dimes. She bought a ticket for 55¢. How much change should she get back?

   ____

3. Pat has 2 quarters. He buys a raffle ticket for 40¢. How much change should he get back?

   ____

4. How many dimes make 90¢?

   ____

5. How many nickels make 65¢?

   ____

Spiral Review

Find each missing addend.

6. $9 + \square = 17$ $\quad 7 + \square = 14$ $\quad 6 + \square = 12$

7. $5 + \square = 11$ $\quad 7 + \square = 13$ $\quad 4 + \square = 12$

Name ________________________________

4•8 Dollars and Cents

Count the money. Write the total amount.

1.

$1.00 $2.00 $3.00

$3.25 $3.35 $3.40 $3.41 $3.42

$ 3 . 42
Total

2.

$ ____ . ____
Total

Problem Solving

3. Rob bought a ticket for 40¢.He gave the clerk two quarters. How much change did Rob get?

____

4. A ticket costs 25¢. Andrew has 3 quarters. How many tickets can he buy?

______ tickets

Spiral Review

Compare. Use >, <, or =.

5.

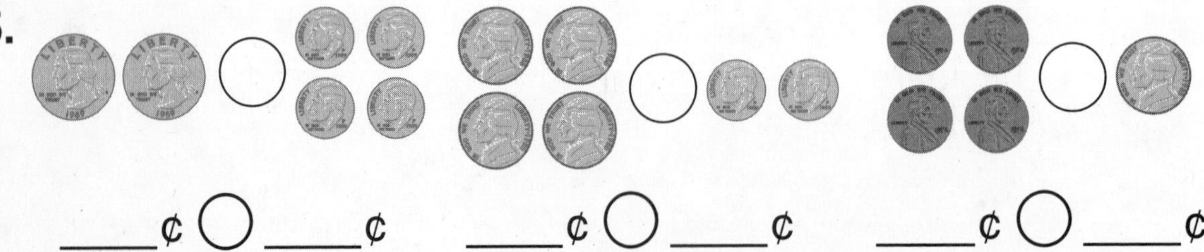

____¢ ◯ ____¢ ____¢ ◯ ____¢ ____¢ ◯ ____¢

Name ________________________________

Compare Money

Count. Is there enough money to buy each item? Circle yes or no.

1. 

  yes no

---

2. 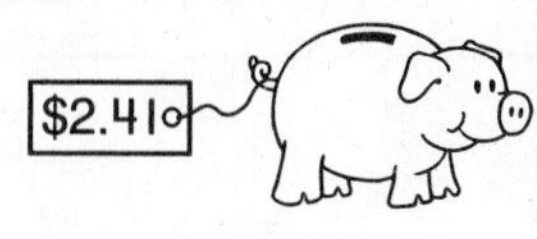

  yes no

---

Problem Solving

3. 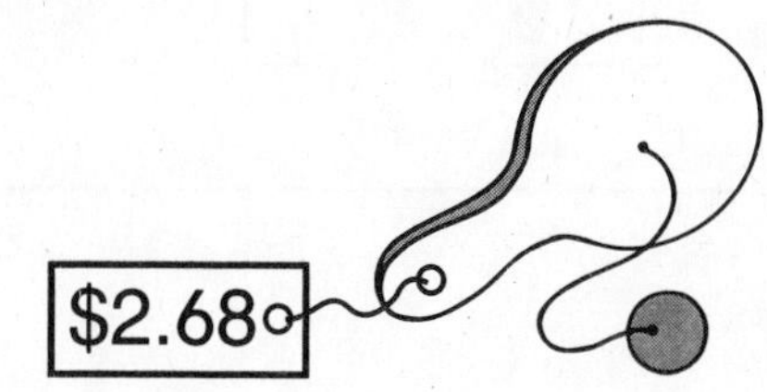

Sara has  . Ellen has . How much do they have altogether? $ ____.____

Is there enough money to buy the item? yes no

Spiral Review

Write the number that comes just before.

4. ______ 63 ______ 35 ______ 79 ______ 40

Write the number that comes just after.

5. 91 ______ 24 ______ 88 ______ 59 ______

Name ______________________________

Add Tens

Add.

1. 42 + 20 = 62　　10 + 36 = ____　　10 + 28 = ____

2. 41 + 30 = ____　　30 + 13 = ____　　20 + 34 = ____

3. 20 + 20　　30 + 34　　26 + 10　　25 + 30　　10 + 12　　17 + 30

4. 26 + 30　　20 + 57　　21 + 20　　30 + 52　　14 + 10　　43 + 10

Problem Solving

5. Fred has 38 nails. Sue has 20 nails. How many nails do they have in all?

____ nails

6. John has 24 nails. Liz has 10 more nails than John. How many nails does Liz have?

____ nails

Spiral Review

Skip count by tens.

7. 10 20 ____ ____ ____ 60 ____ ____ 90

8. 13 23 ____ ____ 53 ____ ____ 83 ____

Name____

2-Digit Addition Without Regrouping

Add.

1.

tens	ones
2 +5	3 +2
7	5

tens	ones
6 +	1 +8

tens	ones
4 +1	2 +5

tens	ones
6 +2	4 +2

2. $31 + 5$ $23 + 4$ $42 + 6$ $53 + 22$ $12 + 37$ $34 + 64$

3. $28 + 51$ $71 + 11$ $35 + 3$ $51 + 25$ $52 + 5$ $21 + 68$

Solve.

4. José made 24 sandwiches. Chris made 12 sandwiches. How many sandwiches did they make all together? ________ sandwiches

5. Ann made 15 glasses of lemonade. Jenna made 13 glasses of lemonade. How many glasses of lemonade in all? ________ glasses

Spiral Review

Write each number in expanded form.

6. 42 ________ 27 ________ 53 ________ 66 ________

Name ____

Decide When to Regroup

Use [tens rod] and [ones cube].

Regroup when you need to.

Add.	Do you need to regroup?	How many in all?
1. 53 + 6	yes (no)	59
2. 46 + 5	yes no	
3. 78 + 4	yes no	
4. 25 + 3	yes no	
5. 57 + 1	yes no	
6. 39 + 2	yes no	

Problem Solving

Find the answer.

7. Becky puts 14 pieces of candy in a jar. Jack adds 8 more pieces. How many pieces of candy all together? ________ pieces

Spiral Review

Count each group of coins.

8.

________¢

9.

________¢

Name________________________________

2-Digit Addition

Decide if you need to regroup.

Add.

1.

tens	ones
1	
2	6
+ 1	5
4	1

tens	ones
4	5
+ 2	3

tens	ones
2	1
+ 6	8

tens	ones
7	3
+	8

2.

tens	ones
1	2
+ 4	9

tens	ones
1	8
+	6

tens	ones
3	4
+	9

tens	ones
5	6
+ 1	7

3.

tens	ones
3	7
+ 1	3

tens	ones
7	9
+	7

tens	ones
2	4
+ 3	5

tens	ones
5	3
+ 4	4

Solve.

4. Jenna makes a pile of 35 bricks. Joe makes a pile of 47 bricks. How many bricks do Jenna and Joe have altogether? ________ bricks

Spiral Review

Add.

5. $\begin{array}{r} 10 \\ +\ 40 \\ \hline \end{array}$ $\begin{array}{r} 23 \\ +\ 50 \\ \hline \end{array}$ $\begin{array}{r} 30 \\ +\ 31 \\ \hline \end{array}$ $\begin{array}{r} 39 \\ +\ 10 \\ \hline \end{array}$ $\begin{array}{r} 65 \\ +\ 20 \\ \hline \end{array}$ $\begin{array}{r} 60 \\ +\ 38 \\ \hline \end{array}$

Name ____________________________________

More 2-Digit Addition

Add. You can use tens and ones models.

1.

tens	ones
☐	
4	2
+	5
4	7

tens	ones
☐	
1	8
+ 2	3

tens	ones
☐	
5	4
+ 3	6

tens	ones
☐	
1	2
+	4

2.

tens	ones
☐	
6	8
+ 1	5

tens	ones
☐	
2	3
+	5

tens	ones
☐	
3	9
+ 3	1

tens	ones
☐	
3	5
+ 2	4

Problem Solving

Solve. Use the chart.

3. How many 🍔 were served on Friday and Saturday in all?

_________ hamburgers

4. Were more 🍔 served on Saturday or Sunday?

____

Served	
Friday	16
Saturday	34
Sunday	22

Spiral Review

Write each number.

5. fifty-five ________ ninety ________ forty-six ________

Name____________________________________

5-6 Practice 2-Digit Addition

Add. You can use tens and ones models.

1.

[1]	[ ]	[ ]	[ ]
27	15	41	58
+ 33	+ 24	+ 4	+ 13
60			

2.

[ ]	[ ]	[ ]	[ ]
13	39	29	26
+ 6	+ 5	+ 12	+ 31

Rewrite. Then add.

3. 32 + 26   [32] + [26] = [58]

48 + 17   [ ] + [ ] = [ ]

55 + 35

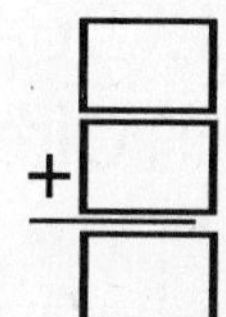

Solve.

4. Myra plants 45 flowers in front of the house.
Jeff plants 25 flowers in back of the house.
How many flowers do they plant all together?
________ flowers

Spiral Review

Circle the coins that show the amount.

5.

6.

Name______________________________________

Problem Solving: Reading for Math

Important and Unimportant Information

10 children visit the zoo. The children see animals that are awake at night. They see 15 bats. They see 16 owls. It is very dark!

How many animals do the children see?

1. Does the information help you to solve the problem? Circle yes or no.

10 children visit the zoo.	yes	(no)
They see 15 bats.	yes	no
They see 16 owls.	yes	no
It is very dark!	yes	no

2. Solve the problem. Write a number sentence.

____

Spiral Review

Add.

3. 15 + 20 = _____ | 33 + 30 = _____ | 27 + 10 = _____

Name ____________________

Problem Solving: Strategy
Draw a Picture

Draw a picture. Solve.

Workspace

1. 10 children gave their reports on Tuesday. 5 children gave their reports on Wednesday. How many children gave reports in all?

__________ children

2. 15 children read a book called *Arthur's Eyes.* 8 children read a book called *Charlotte's Web.* How many children read books?

__________ children

3. 11 children will bring their lunches to school. 7 children will buy their lunch. How many children will eat lunch all together?

__________ children

Spiral Review

Add.

4.

tens	ones
□	
1	8
+2	7

tens	ones
□	
3	3
+4	6

tens	ones
□	
5	2
+3	4

tens	ones
□	
8	5
+	8

Name ____________________

Check Addition

Add. Check by adding in a different order.

1.
$$\begin{array}{r} 43 \\ +\ 25 \\ \hline 68 \end{array} \quad \begin{array}{r} 25 \\ +\ 43 \\ \hline 68 \end{array} \qquad \begin{array}{r} 25 \\ +\ 62 \\ \hline \end{array} \quad + \underline{\quad} \qquad \begin{array}{r} 33 \\ +\ 49 \\ \hline \end{array} \quad + \underline{\quad}$$

2.
$$\begin{array}{r} 6 \\ +\ 42 \\ \hline \end{array} \quad + \underline{\quad} \qquad \begin{array}{r} 21 \\ +\ 49 \\ \hline \end{array} \quad + \underline{\quad} \qquad \begin{array}{r} 52 \\ +\ 36 \\ \hline \end{array} \quad + \underline{\quad}$$

3.
$$\begin{array}{r} 9 \\ +\ 39 \\ \hline \end{array} \quad + \underline{\quad} \qquad \begin{array}{r} 44 \\ +\ 13 \\ \hline \end{array} \quad + \underline{\quad} \qquad \begin{array}{r} 13 \\ +\ 38 \\ \hline \end{array} \quad + \underline{\quad}$$

Solve. Use the chart. Check by adding in a different order.

4. Which two children sold the most cookies?

____

How many boxes did these two children sell in all?

________ boxes

Cookies Sold

Gayle	39 boxes
Joy	26 boxes
Keith	25 boxes
Matt	43 boxes
Kate	54 boxes

Spiral Review

Compare. Use >, <, or =.

5. 83 ◯ 76   45 ◯ 45   29 ◯ 31   6 ◯ 10

Name ______________________________

Estimate Sums

Add. Estimate to see if your answer is reasonable.

1. $\begin{array}{r} 51 \\ +28 \\ \hline 79 \end{array}$ $\begin{array}{r} 50 \\ +30 \\ \hline 80 \end{array}$	$\begin{array}{r} 29 \\ +32 \\ \hline \end{array}$ $\begin{array}{r} \\ +\_\_ \\ \end{array}$	$\begin{array}{r} 31 \\ +28 \\ \hline \end{array}$ $\begin{array}{r} \\ +\_\_ \\ \end{array}$
2. $\begin{array}{r} 22 \\ +49 \\ \hline \end{array}$ $\begin{array}{r} \\ +\_\_ \\ \end{array}$	$\begin{array}{r} 13 \\ +27 \\ \hline \end{array}$ $\begin{array}{r} \\ +\_\_ \\ \end{array}$	$\begin{array}{r} 62 \\ +17 \\ \hline \end{array}$ $\begin{array}{r} \\ +\_\_ \\ \end{array}$
3. $\begin{array}{r} 57 \\ +29 \\ \hline \end{array}$ $\begin{array}{r} \\ +\_\_ \\ \end{array}$	$\begin{array}{r} 38 \\ +49 \\ \hline \end{array}$ $\begin{array}{r} \\ +\_\_ \\ \end{array}$	$\begin{array}{r} 41 \\ +48 \\ \hline \end{array}$ $\begin{array}{r} \\ +\_\_ \\ \end{array}$

Problem Solving

Solve. Estimate to see if your answer is reasonable.

4. Lisa spends 21 minutes on math homework and 18 minutes on spelling homework. How much time does she spend on homework in all?

__________ minutes

5. Becky's class takes 49 books out of the library. Barry's class takes out 37 books. How many books do the two classes take out in all?

__________ books

Spiral Review

Write the number that comes just after.

6. 34 [ ]   47 [ ]   83 [ ]   22 [ ]

Name ______________________________

Add Money Amounts

Add.

1.	65¢ + 8¢ 72¢	48¢ + 38¢	13¢ + 44¢	22¢ + 35¢	41¢ + 8¢	65¢ + 6¢
2.	53¢ + 15¢	11¢ + 16¢	24¢ + 9¢	32¢ + 47¢	14¢ + 58¢	25¢ + 38¢
3.	26¢ + 36¢	23¢ + 12¢	49¢ + 26¢	17¢ + 13¢	37¢ + 8¢	33¢ + 28¢

Problem Solving

Solve.

4. Keisha buys 2 charms for 25¢ each. How much does she spend in all? __________

Spiral Review

Add.

6.	8 4 + 2	8 7 + 3	9 9 + 1	6 4 + 7	9 5 + 1	6 5 + 5

Name____

Three Addends

Add. You can use tens and ones models.

1.
25	32	38	56	25	4
18	22	8	21	12	77
+ 35	+ 14	+ 13	+ 2	+ 45	+ 17
78					

2.
17	22	37	22	28	29
6	19	23	43	30	10
+ 34	+ 25	+ 26	+ 32	+ 21	+ 39

Problem Solving

Solve. Use the chart.

Grade 2 Favorite Colors

blue	31 votes
red	22 votes
purple	15 votes
green	14 votes
yellow	6 votes
orange	2 votes

3. Which 3 colors got the most votes?

____

How many votes did these colors get in all?

_________ votes

4. Which 3 colors got the fewest votes? _______________

How many votes did these colors get in all? _________ votes

Spiral Review

Subtract.

5.
18	13	16	17	14	15
− 9	− 7	− 8	− 9	− 8	− 9

Name ______________________________

Subtract Tens

Remember to count back by tens.

Subtract.

1. 86 − 30 = ______ 92 − 20 = ______ 47 − 20 = ______

2. 84 − 20 = ______ 61 − 10 = ______ 38 − 30 = ______

3.

53	93	30	78	52	38
− 30	− 10	− 20	− 10	− 30	− 20

4.

77	35	81	49	87	89
− 30	− 20	− 10	− 30	− 20	− 10

Solve.

5. 54 people wait in line for movie tickets.
20 people buy tickets and go inside.
How many people are left in the ticket line? ______ people

6. Linda has 36 bags of popcorn. She sells
20 bags of popcorn. How many bags are left? ______ bags

Spiral Review

Count up to find the change.

	Ticket Price	You Pay	Change
7.			______ change

Name________________________________

2-Digit Subtraction Without Regrouping

Subtract.

1.

	tens	ones
	3	5
−	2	1
	1	4

	tens	ones
	8	9
−	4	2

	tens	ones
	6	6
−	6	3

2. $75 - 60$ $22 - 11$ $93 - 2$ $64 - 31$ $86 - 54$ $59 - 7$

3. $58 - 7$ $59 - 17$ $66 - 12$ $57 - 43$ $94 - 72$ $88 - 67$

4. $56 - 2$ $97 - 32$ $79 - 11$ $76 - 4$ $68 - 15$ $87 - 3$

Solve.

5. 47 people are on the bus. 25 people get off at the first stop. How many people are left on the bus? ______ people

Spiral Review

Add.

6. $53 + 27$ $29 + 55$ $56 + 43$ $45 + 14$ $38 + 35$ $79 + 13$

Name ______________________________

Decide When to Regroup

	Subtract	Do you need to regroup?	How many are left?
1.	68 – 9	yes no	59
2.	45 – 3	yes no	______
3.	77 – 5	yes no	______
4.	94 – 6	yes no	______
5.	52 – 4	yes no	______
6.	39 – 7	yes no	______

Problem Solving

7. Marc works in the pet store.
He has 36 birds to sell.
People buy 8 birds.
How many birds are left? ______ birds

8. Marc has 24 boxes of bird seed.
He sells 3 boxes of bird seed.
How many boxes of bird seed are left? ______ boxes

Spiral Review

Complete each number sentence. Use + or – .

9. 13 ◯ 5 = 8   8 ◯ 4 = 12   9 ◯ 2 = 11

10. 17 ◯ 9 = 8   9 ◯ 6 = 15   15 ◯ 8 = 7

Name ______________________________

2-Digit Subtraction

Subtract. You can use tens and ones models.

1.

tens	ones
3	1
− 1	1
2	0

tens	ones
4	1
− 1	2

tens	ones
5	6
− 1	4

tens	ones
6	6
− 1	7

2.

tens	ones
4	5
− 1	8

tens	ones
9	8
−	5

tens	ones
9	2
− 4	6

tens	ones
3	6
− 1	7

3.

tens	ones
7	8
− 3	5

tens	ones
7	8
− 3	9

tens	ones
4	3
−	8

tens	ones
3	2
− 2	8

Solve.

4. Jan counts 35 birds. 18 are robins. How many birds are not robins? ______ birds

5. The second graders have a pet fair. 25 pets are shown. 3 pets are hamsters. How many are not hamsters? ______ pets

Spiral Review

Add. Estimate to see if your answer is reasonable.

6.

33	30
+ 18	+ 20
	50

52	
+ 27	+ ____

49	
+ 21	+ ____

19	
+ 31	+ ____

Name ____________________

More 2-Digit Subtraction

Subtract. You can use tens and ones models.

1.

	tens	ones
	☐	☐
	5	8
−		5

	tens	ones
	☐	☐
	6	1
−	1	5

	tens	ones
	☐	☐
	8	6
−	2	1

	tens	ones
	☐	☐
	8	1
−	2	2

2.

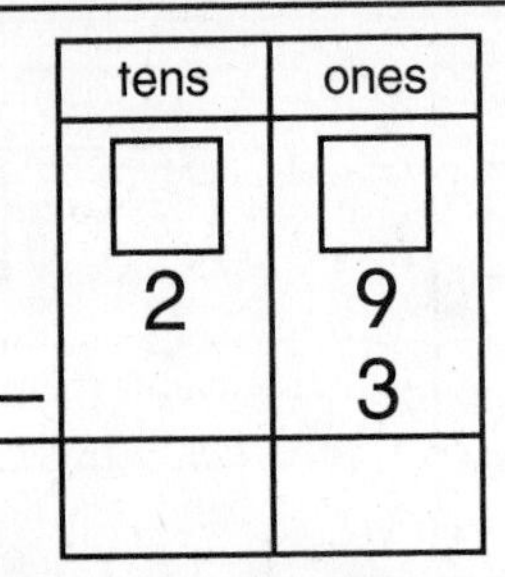

	tens	ones
	☐	☐
	2	9
−		3

	tens	ones
	☐	☐
	3	3
−	1	7

	tens	ones
	☐	☐
	4	3
−	1	7

	tens	ones
	☐	☐
	3	4
−	1	7

Problem Solving

Solve. Use the chart.

4. Swimmers are in 3 groups. How many more swimmers are in the Tadpoles group than the Minnows group?

______ campers

5. How many more swimmers are in the Minnows group than the Sunfish group? campers ______

Groups for Swim Class

Group	Campers
Tadpoles	55 campers
Minnows	43 campers
Sunfish	39 campers

Spiral Review

Count. Is there enough money to buy the bucket? Circle yes or no.

5. 

  ______ yes no

Name ______________________________

Practice 2-Digit Subtraction

Subtract. You can use tens and ones models.

1.

5 13					
~~6~~ ~~3~~	75	34	87	55	41
− 37	− 22	− 16	− 78	− 3	− 23
26					

2.

77	59	84	99	64	57
− 45	− 7	− 29	− 38	− 39	− 9

3.

91	68	92	65	74	94
− 49	− 36	− 56	− 8	− 31	− 19

Solve.

4. Jim's team scores 58 points. Eve's team scores 65 points. By how many points does Eve's team win?

   ____________ points

Spiral Review

Write as tens and ones. Then write in expanded form.

6. 74 = __7__ tens __4__ ones = __70__ + __4__

7. 93 = ______ tens ______ ones = ______ + ______

8. 85 = ______ tens ______ ones = ______ + ______

Name ____________________

Problem Solving: Reading for Math
Steps in a Process

Answer each question.

Ella and Bert work in the garden on Monday. First, they pick 24 tomatoes. Next, they pick 15 peppers. After that, they rinse the dirt off the tomatoes and peppers. Finally, they put the tomatoes and peppers in baskets.

1. What do Ella and Bert do first? ____________________

2. What do they do next? ____________________

3. What do they do after that? ____________________

4. What did they do last? ____________________

5. Write a subtraction problem about the story. ____________________.

____

Spiral Review

What could the next number be? Find the pattern.

6. 77 67 57 47 ______ 43 33 23 13 ______

Name ___________________________________

Problem Solving Strategy:
Choose the Operation

Circle add or subtract. Solve.

1. 10 children go to the library. 8 more children join them. How many children are in the library altogether?

   add subtract

   ____

   ______ children

2. 16 children read picture books. 9 read chapter books. How many more read picture books?

   add subtract

   ____

   ______ children

3. Ms. Lee's class reads 27 books. Mr. Brown's class 34. How many books do they read in all?

   add subtract

   ____

   ______ books

4. Students take 45 books out of the library. 23 are returned. How many have not been returned?

   add subtract

   ____

   ______ books

Spiral Review

Complete each fact family.

5. 8 + 2 = ______	9 + 8 = ______	______ + 4 = 11
2 + 8 = ______	8 + 9 = ______	4 ◯ 7 = 11
10 − ______ = 8	17 ◯ 9 = 8	______ − 7 = 4
______ − 8 = 2	17 − ______ = 9	11 − ______ = 7

Name ____________________

6·9 Check Subtraction

Subtract. Check by adding.

1.

$$\begin{array}{r} 86 \\ -\ 44 \\ \hline 42 \end{array} \quad \begin{array}{r} 42 \\ +\ 44 \\ \hline 86 \end{array}$$

$$\begin{array}{r} 83 \\ -\ 69 \\ \hline \end{array} \quad + \underline{\quad}$$

$$\begin{array}{r} 72 \\ -\ 48 \\ \hline \end{array} \quad + \underline{\quad}$$

$$\begin{array}{r} 35 \\ -\ 17 \\ \hline \end{array} \quad + \underline{\quad}$$

2.

$$\begin{array}{r} 62 \\ -\ 26 \\ \hline \end{array} \quad + \underline{\quad}$$

$$\begin{array}{r} 95 \\ -\ 13 \\ \hline \end{array} \quad + \underline{\quad}$$

$$\begin{array}{r} 96 \\ -\ 79 \\ \hline \end{array} \quad + \underline{\quad}$$

$$\begin{array}{r} 61 \\ -\ 43 \\ \hline \end{array} \quad + \underline{\quad}$$

3.

$$\begin{array}{r} 81 \\ -\ 32 \\ \hline \end{array} \quad + \underline{\quad}$$

$$\begin{array}{r} 54 \\ -\ 28 \\ \hline \end{array} \quad + \underline{\quad}$$

$$\begin{array}{r} 82 \\ -\ 21 \\ \hline \end{array} \quad + \underline{\quad}$$

$$\begin{array}{r} 42 \\ -\ 27 \\ \hline \end{array} \quad + \underline{\quad}$$

Solve. Use the chart. Check by adding.

Reading Club

Name	Books
Ezra	92 books
Kathy	85 books
Mike	71 books

4. How many more books does Ezra read than Kathy?

_____ books

5. How many more books does Kathy read than Mike?

_____ books

Spiral Review

Add.

6.

$$\begin{array}{r} 16 \\ 34 \\ +\ 23 \\ \hline \end{array} \quad \begin{array}{r} 23 \\ 23 \\ +\ 12 \\ \hline \end{array} \quad \begin{array}{r} 22 \\ 47 \\ +\ 13 \\ \hline \end{array} \quad \begin{array}{r} 27 \\ 21 \\ +\ 31 \\ \hline \end{array} \quad \begin{array}{r} 45 \\ 24 \\ +\ 25 \\ \hline \end{array} \quad \begin{array}{r} 37 \\ 12 \\ +\ 28 \\ \hline \end{array}$$

Name ______________________________

6-10 Estimate Differences

Subtract. Estimate to see if your answer is reasonable.

1.	29 − 18 = 11; 30 − 20 = 10	31 − 12 ; − ___	92 − 22 ; − ___	88 − 29 ; − ___
2.	51 − 22 ; − ___	89 − 32 ; − ___	79 − 21 ; − ___	82 − 28 ; − ___
3.	77 − 18 ; − ___	36 − 19 ; − ___	64 − 16 ; − ___	73 − 59 ; − ___

Problem Solving

Solve. Estimate to see if your answer is reasonable.

4. Janelle collects 22 newspapers for recycling. Erin collects 19 newspapers. How many more newspapers does Janelle collect than Erin?

_____ newspapers

5. Eric collects 48 bottles and cans. 31 are bottles. How many are cans?

_____ cans

Spiral Review

Add or subtract.

7. 25 − 16   33 + 49   26 + 32   61 − 28   53 + 7   46 + 39

Name ____________________

Subtract Money Amounts

Subtract. Remember to write the ¢ sign.

1.

63¢	98¢	87¢	68¢	54¢	38¢
− 45¢	− 88¢	− 34¢	− 23¢	− 47¢	− 17¢
18¢					

2.

26¢	76¢	58¢	74¢	49¢	26¢
− 18¢	− 38¢	− 39¢	− 15¢	− 42¢	− 7¢

3.

52¢	34¢	73¢	84¢	29¢	68¢
− 41¢	− 7¢	− 68¢	− 23¢	− 18¢	− 49¢

4.

64¢	14¢	96¢	83¢	44¢	82¢
− 18¢	− 11¢	− 47¢	− 38¢	− 23¢	− 72¢

Problem Solving

5. Kitty buys an apple for 37¢. She gives the clerk 50¢. She gets back 15¢ change. Is this correct? Why or why not?

____

6. Tray buys a pear for 63¢. He gives the clerk 75¢. How much change does he receive? ________

Spiral Review

Complete each number sentence. Use + or −.

7. 19 ◯ 5 = 14   6 ◯ 4 = 10   9 ◯ 8 = 17   16 ◯ 7 = 9

Name________________________________

Time to the Hour and Half Hour

Write each time.

Draw the minute hand to show each time.

Spiral Review

Brad and Janelle like to travel. Brad has visited 9 states. Janelle has visited 14 states. How many more states has Janelle visited than Brad?

Does the information help you to solve the problem? Circle yes or no.

3. Brad and Janelle like to travel.	yes	no
4. Brad has visited 9 states.	yes	no
5. Janelle has visited 14 states.	yes	no

6. Solve the problem. Write a number sentence.

____

Name ____

7-2 Time to Five Minutes

Write the time for each clock.

1. 			
8:00	:	:	:
2. 			
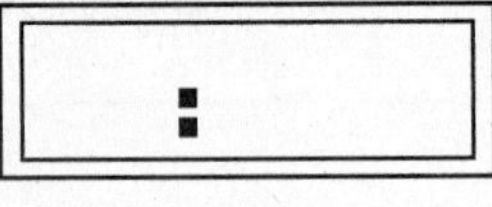 :	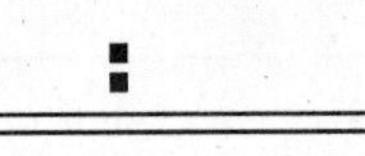 :	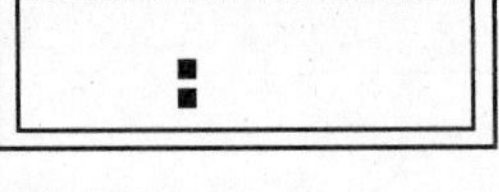 :	:

Solve.

3. 

The bus arrives at the stop at 6:20.
The next bus arrives in 5 minutes.
What time does it arrive?

____

Spiral Review

Add.

4.

34	28	63	25	55	22
+ 52	+ 46	+ 19	+ 26	+ 13	+ 49

Name________________________________

7·3 Time to the Quarter Hour

Write the time for each clock.

1.			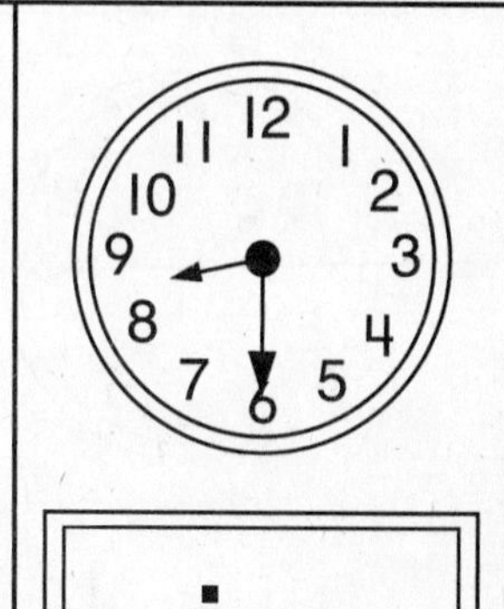
7 : 45	:	:	:

Draw the minute hand to show each time.

2. 2 : 00	2 : 15	2 : 30

Solve.

3.

Recess begins at 10:00.
It ends 15 minutes later.
What time does it end?

____

Spiral Review

Write each number in words.

4. 15 ______________ 60 ______________ 35 ______________

Name ____________________

7·4 More Time

Write each time.

1.  10 : 15 quarter after 10	___ : ___ ___ minutes after ___ ___ minutes before ___

Problem Solving

Draw hour and minute hands to show each time.

2.

8 :22

4 :01

10:17

5 :52

Spiral Review

Subtract. Estimate to see if your answer is reasonable.

3.

$$\begin{array}{r} 88 \\ -41 \\ \hline \end{array} \qquad \begin{array}{r} 82 \\ -23 \\ \hline \end{array} \qquad \begin{array}{r} 61 \\ -17 \\ \hline \end{array}$$

Name ______________________________

7•5 Problem Solving: Reading for Math

Sequence of Events

Solve.

Dear Jill:
You will love camp! We get up at 8:30. Swim lessons are at quarter to 11. Lunch is at 20 minutes after 12. My favorite thing is the campfire at 7:00 each night.

Love,
Debra

1. What time do the campers wake up?

____

2. At what time do swim lessons begin? Write the time 3 ways.

____

____

3. When is lunch? _______________

4. What happens at 7:00? _______________

Spiral Review

Count to find the total amount.
Use >, <, or = to complete the sentence.

5. 

_______________ ◯ _______________

_______________ ◯ _______________

Name ____________________

Problem Solving: Strategy
Act It Out

Use a clock to act it out. Solve.

1. It takes a half hour to walk from Kate's house to the Children's Museum. How long does it take to walk there and back?

____

2. It takes 15 minutes to see each part of the museum. How long does it take to see 3 parts?

____

3. The museum has a sing-along. It takes 20 minutes. How long would 3 sing-alongs take?

____

4. Bill listens to tapes at the museum. Each tape is 5 minutes long. How long does it take to listen to 2 tapes?

____

5. Amy makes a stick puppet in 20 minutes. How long would it take to make 2 puppets?

____

6. Marie watches 3 movies at the museum. Each is 20 minutes long. How long does it take to watch all 3 movies?

____

Spiral Review

Skip count by threes.

7. 3 ____ 9 ____ ____ 18 ____ ____

Skip count by fives.

8. 5 10 ____ ____ ____ 30 ____ ____

Name ______________________________

Elapsed Time

Complete the table.

1.	Start time	End time	Elapsed time
 Watch a movie.	 ____ : ____	 ____ : ____	____ hours

Draw the clock hands to show the elapsed time.

2.		
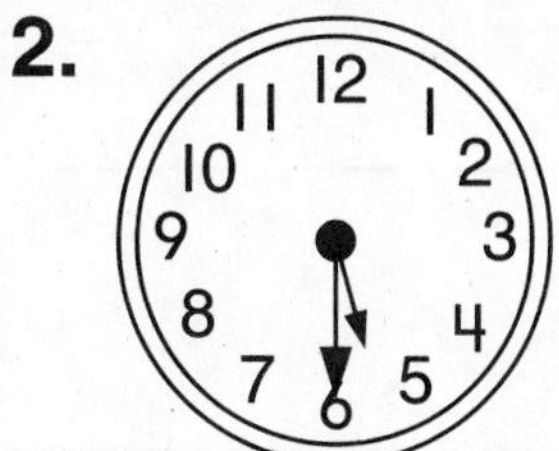	2 hours later	

Spiral Review

Write the time for each clock.

3.  

______________ ______________ ______________ ______________

Name ______________________________

Calendar

January 2002						
S	M	T	W	T	F	S
		1	2	3	4	5
6	7	8	9	10	11	12
13	14	15	16	17	18	19
20	21	22	23	24	25	26
27	28	29	30	31		

February 2002						
S	M	T	W	T	F	S
					1	2
3	4	5	6	7	8	9
10	11	12	13	14	15	16
17	18	19	20	21	22	23
24	25	26	27	28		

March 2002						
S	M	T	W	T	F	S
					1	2
3	4	5	6	7	8	9
10	11	12	13	14	15	16
17	18	19	20	21	22	23
24	25	26	27	28	29	30
31						

April 2002						
S	M	T	W	T	F	S
	1	2	3	4	5	6
7	8	9	10	11	12	13
14	15	16	17	18	19	20
21	22	23	24	25	26	27
28	29	30				

Use the calendar to answer each question.

1. How many days are there between February 14 and February 25? __________ days

2. What day of the week is January 1? __________

3. What month is just before April? __________

4. What is the date of the third Wednesday in March? __________

Problem Solving

5. Today is April 20. Anna's party is April 27. How many days until her party? __________ days

Spiral Review

Draw the minute hand to show each time.

7.

8:15

10:30

3:45

Name ____________________

Read Pictographs

Use the pictograph. Answer each question.

Second Graders' Favorite Foods

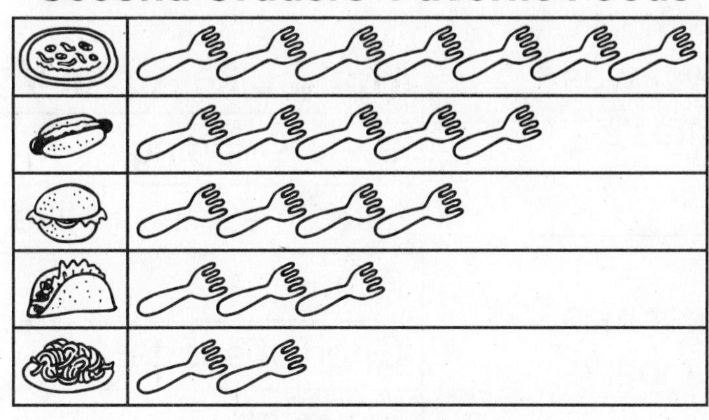

1 [fork] = 2 votes

1. How many votes does each [fork] stand for? ________ votes
2. How many children voted? ________ children
3. How many children voted for pizza? ________ children
4. How many more children voted for pizza than burgers? ________ children

Problem Solving

5. What if each [fork] stands for 5 votes. How many children voted for tacos? ________ children

Spiral Review

Add.	Do you need to regroup?		How many in all?
6. 46 + 11	yes	no	________
7. 58 + 33	yes	no	________

Name ______________________________

Tally Marks and Charts

Becca took a survey. She asked her classmates what they like about their best friends.

What I Like About My Best Friend

Understanding	𝍸 ǁ
Likes the things I like	𝍸 ǀ
Funny	𝍸
Good at sports	ǀǀǀǀ
Plays fair	ǀǀǀ
Shares	ǁ

1. Which answer was chosen 5 times?

____

2. Which answer was chosen the most?

____

3. Which answer was chosen the fewest times? ______________

4. How many students did Becca ask? ______________ students

5. How many students chose "Plays fair" and "Shares"?

______________ students

6. How many more students chose "Understanding" than "Good at sports"? ______________ students

Spiral Review

Add. Check by adding in a different order.

7.

24		38		11	
+ 37	+ ___	+ 56	+ ___	+ 44	+ ___

Name ______________________

Bar Graphs

Use the totals to complete the bar graph. Color one space to show each vote.

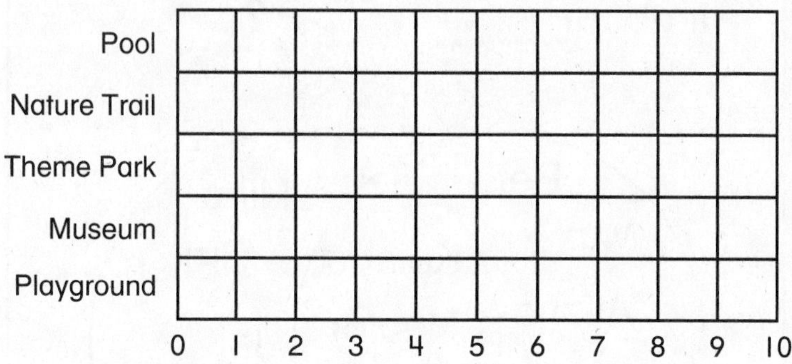

Use data from the graph to answer each question.

1. Which place did the greatest number of children vote for? ______________

2. How many places does the graph show? ________ places

3. How many children voted in all? ________ children

4. How many more children voted for the museum than for the pool? ________ children

Spiral Review

Write each number.

5. forty-eight ________ twenty-one ________ sixty ________

6. seventy-six ________ fifty-two ________ seventeen ________

Name ______________________________

Problem Solving: Reading for Math
Draw Conclusions

Hector collected data about his second-grade class.

Do you like to learn about the planets or about dinosaurs?

like planets	like both	like dinosaurs
Beth	Keisha	Nina
Max	Hector	Ellen
Alisha	Kate	Jeff
Kendra	Carl	Tim
Josh		Ed
Jim		

1. How many children like planets and dinosaurs?

   __4__ children

2. How many children like dinosaurs?

   ________ children

3. How many children like planets but not dinosaurs? ________ children

4. Would more children rather learn about the planets or about dinosaurs? Explain your thinking.

   ____

   ____

Spiral Review

Write each as a subtraction sentence. Complete.

5. $\square + 23 = 67$ $\square + 28 = 54$ $\square + 19 = 55$

   ___ − ___ = ___   ___ − ___ = ___   ___ − ___ = ___

Name______________________________

Problem Solving: Strategy
Make a Table

Fill in the table. Then answer the questions.

Tim's class sold 12 boxes of mint cookies and 10 boxes of chip cookies. They also sold 14 boxes of nut cookies and 9 boxes of oatmeal cookies.

Boxes of Cookies Sold

Kind	Mint	Chip	Nut	Oatmeal
How Many?				

1. What kind of cookie sold the fewest? ______________

2. How many more boxes of nut cookies were sold

   than oatmeal cookies? ________ boxes

3. How many boxes of mint cookies and nut cookies

   were sold in all? ________ boxes

4. How could you find out how many boxes of cookies were sold in all?

   ____

Spiral Review

Add or subtract.

5. $\begin{array}{r} 64 \\ -\ 36 \\ \hline \end{array}$ $\begin{array}{r} 22 \\ +\ 65 \\ \hline \end{array}$ $\begin{array}{r} 34 \\ +\ 47 \\ \hline \end{array}$ $\begin{array}{r} 72 \\ -\ 49 \\ \hline \end{array}$ $\begin{array}{r} 55 \\ -\ 31 \\ \hline \end{array}$ $\begin{array}{r} 43 \\ +\ 48 \\ \hline \end{array}$

Name ____

8-6 Represent Data in Different Ways

Use the data from the tally table to make the bar graph. Answer each question.

Favorite Sport

Soccer	~~IIII~~ II
Swimming	~~IIII~~ I
Softball	IIII

1. Which sport got 4 votes?

____

Favorite Sport

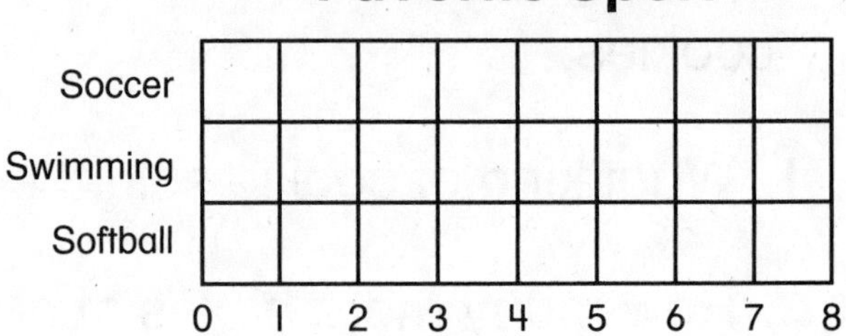

2. From which data display did you get your answer for question 1? Explain.

____

____

____

3. Which sport got the most votes? ____________________________

4. From which data display did you get your answer for question 3? Explain.

____

____

Spiral Review

5. Matt has 1 quarter, 2 dimes, and 2 pennies. His brother gives him 2 more pennies. How much money does Matt have now? ______________

Name ____________________

8•7 Range and Mode

Use the data. Answer each question.
The following numbers show how many baseball caps the school store sold in the past five weeks.

5 3 10 24 5

1. What is the range of the numbers? ____________
2. What is the mode of the numbers? ____________

The following numbers show how many pencils the school store sold in the last five weeks.

25 36 18 25 16

3. What is the range of the numbers? ____________
4. What is the mode of the numbers? ____________

Spiral Review

5. 

John began watching a movie at 1:00. The movie ended at 3:00. How long did he spend watching the movie?

____________ hours

6. 

Carla started playing soccer at 2:00. She played for 1 hour and 30 minutes. At what time did she finish playing soccer?

________ : ________

Name ________________________________

Inch and Foot

Find these objects.
Estimate. Then measure and record.
Write how many inches or feet.

	Estimate	Measure
1.	about ______	______
2.	about ______	______
3.	about ______	______

Solve.

4. A piece of notebook paper is about 9 inches wide. How wide are two pieces of paper taped side by side? ______ inches

Spiral Review

Add or subtract. Complete the fact family.

5. 6 + 7 = ______ 13 − 6 = ______

7 + ______ = 13 ______ − 7 = 6

Name ____________________

Inches, Feet, and Yards

Find these objects.
Estimate. Then measure and record. Write how many inches, feet, or yards.

	Estimate	Measure
1.	about ______	______
2.	about ______	______
3. THE NEWS	about ______	______

Solve.

4. The second graders are having a relay race. Beth runs 15 yards. She passes the ball to Jack, who runs 16 more yards. How many yards do they run in all? ______ yards

Spiral Review

Add.

5.

22	33	18	17	6	52
38	23	26	13	32	24
+ 15	+ 11	+ 34	+ 44	+ 28	+ 12

Name ______________________________

Cup, Pint, Quart

Circle the better estimate.

1. 

more than 1 quart

less than 1 quart

2. 

more than 1 quart

less than 1 quart

3. 

more than 1 pint

less than 1 pint

4. 

more than 1 cup

less than 1 cup

Write the answer.

5. Jamie drinks a cup of water at lunch. He drinks a pint of water at dinner. Does he drink more water at lunch or at dinner? ____________

6. Rich uses 4 quarts of water to fill the fish bowl. He uses 3 quarts to water the plants. How many quarts of water does he use in all? ______ quarts

Spiral Review

Add.

7.					
67	43	21	39	46	19
+ 25	+ 14	+ 27	+ 42	+ 29	+ 35

Name ____________________

9•4 Ounce and Pound

Circle the better estimate.

1.  

more than 1 ounce

less than 1 ounce

more than 1 pound

less than 1 pound

2. 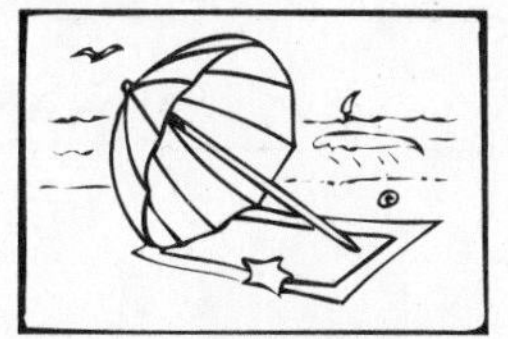

more than 1 ounce

less than 1 ounce

more than 1 pound

less than 1 pound

Solve.

3. Cindy and Max weigh some objects. A large shell weighs 18 ounces. A rock weighs 1 pound. Which object weighs more? Explain.

____

____

Spiral Review

Subtract.

4.

74	55	89	62	94	85
− 48	− 36	− 53	− 47	− 41	− 37

Name________________________________

Perimeter

Find the perimeter of each figure.
Use an inch ruler.

1.
1 in.
1 in.
1 in.
1 in.

1 + 1 + 1 + 1 = 4 inches

2.
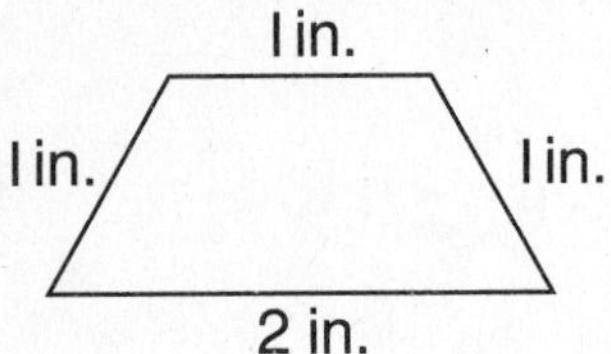

____ + ____ + ____ + ____ = ____ inches

3.
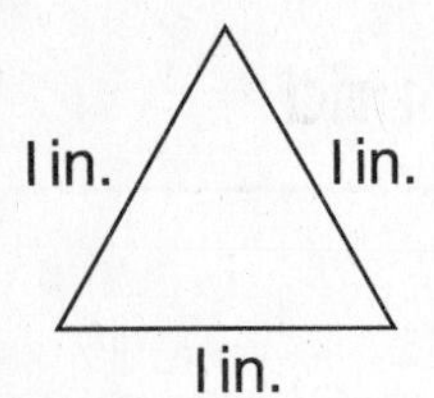

____ + ____ + ____ = ____ inches

Problem Solving

4.
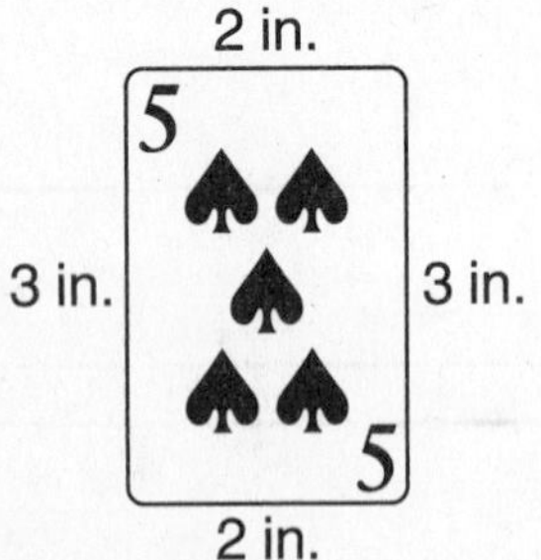

A playing card is 2 inches wide. The card is 3 inches tall. What is the perimeter?

___ + ___ + ___ + ___ = ___ inches

Spiral Review

Circle the better estimate.

5. 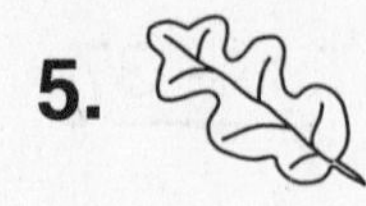
more than 1 pound
less than 1 pound

more than 1 ounce
less than 1 ounce

Name ______________________________

Area

Color to show the number of square units.

1.

6 square units

8 square units

2.

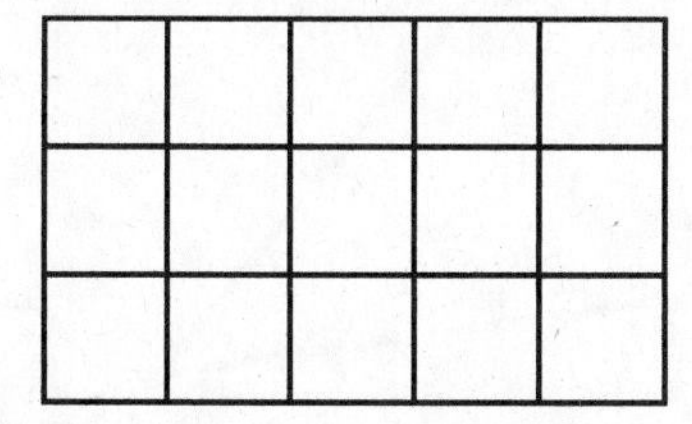

9 square units

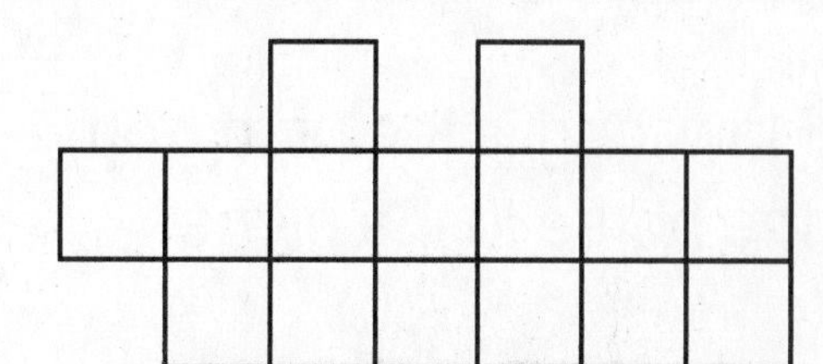

11 square units

Solve.

3. The area of a piece of Nancy's notebook paper is 16 square units. What is the area of two pieces of notebook paper? ______ square units

4. The area of Joe's placemat is 24 square units. The area of his napkin is 16 square units. How many more square units is the placemat than the napkin? ______ square units

Spiral Review

Count on the coins. Find the change.

5.

Change: ______ ¢

Name____

9·7 Problem Solving: Reading for Math

Use Maps

Use your ruler to measure.

Solve.

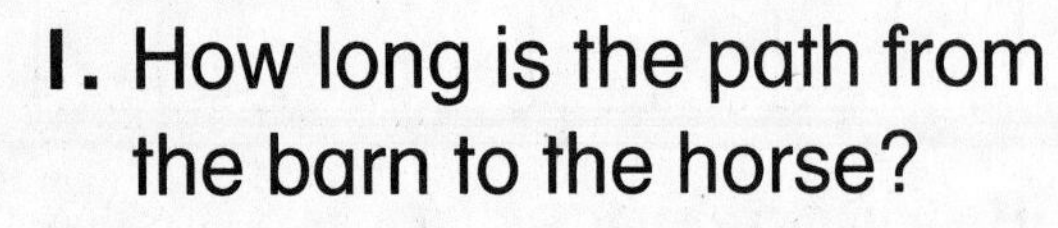

1. How long is the path from the barn to the horse?

   ____2____ inches

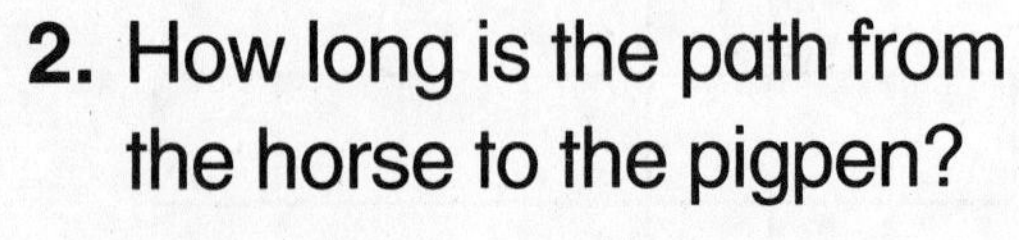

2. How long is the path from the horse to the pigpen?

   ________ inch

3. How long is the path from the apple trees to the pigpen? ________ inches

4. How long is the path from the barn to the apple trees? ________ inches

5. Ed moves a paper clip from the barn to the apple trees. Then he moves the paper clip from the apple trees to the pigpen. How far does he move the paper clip? ________ inches

6. What is the shortest path from the barn to the pigpen? Explain.

____

____

Spiral Review

Complete the sentence. Use >, <, or =.

7. 14 inches ◯ 1 foot  3 feet ◯ 1 yard  1 quart ◯ 5 cups

Name ____________________

Problem Solving: Strategy

Guess and Check

How long is each path? Estimate the length. Then use a ruler to check.

1.

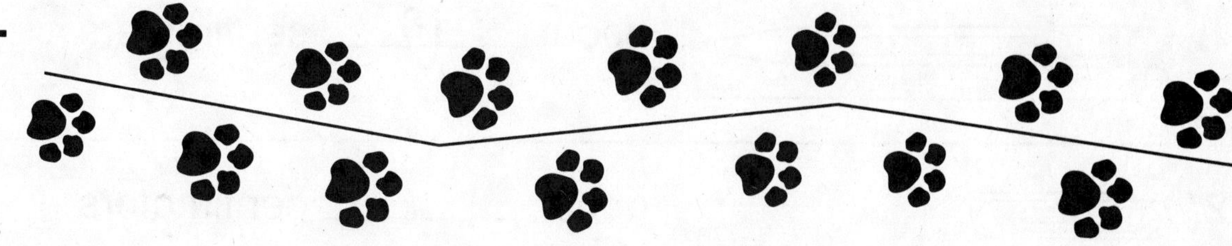

Estimate ___6___ inches Check ___6___ inches

2.

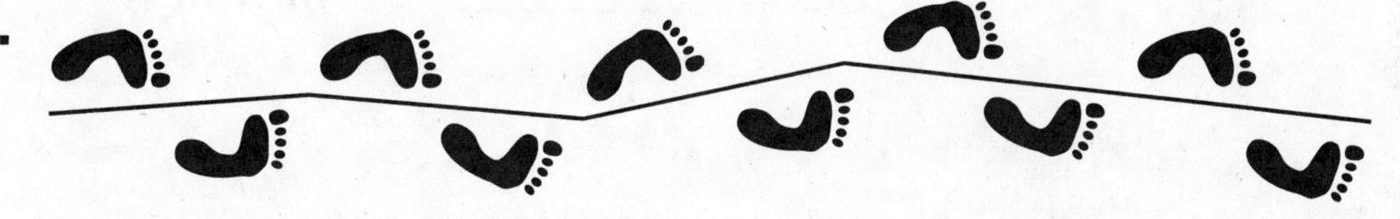

Estimate ______ inches Check ______ inches

3.

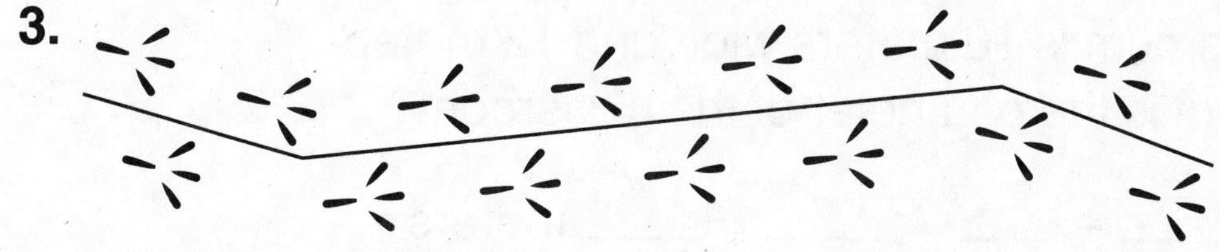

Estimate ______ inches Check ______ inches

Spiral Review

Write the time shown.

4.

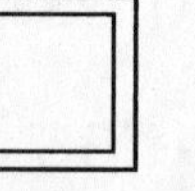 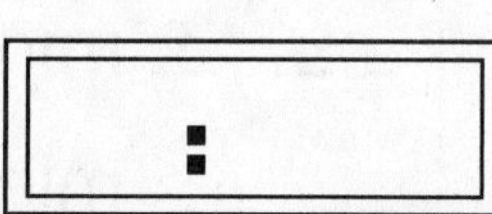 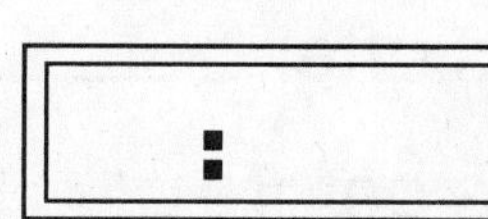

: : : :

Name ______________________________

Centimeter and Meter

Find these objects.
Use a centimeter ruler to measure.

1. about ___18___ centimeters

2. about ________ centimeters

3. about ________ centimeters

Solve.

4. The classroom is 10 meters wide and 12 meters long. What is the perimeter of the classroom?

_____ + _____ + _____ + _____ = _____ meters

Spiral Review

5. Write each time three ways.

______ : ______

______ minutes after ______

______ minutes before ______

______ : ______

______ minutes after ______

______ minutes before ______

Name________________________________

Gram and Kilogram

Choose the better estimate.

1.

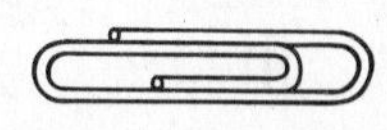

lighter than 1 kilogram

heavier than 1 kilogram

about the same as 1 kilogram

lighter than 1 kilogram

heavier than 1 kilogram

about the same as 1 kilogram

2.

lighter than 1 kilogram

heavier than 1 kilogram

about the same as 1 kilogram

lighter than 1 kilogram

heavier than 1 kilogram

about the same as 1 kilogram

Problem Solving

3. Martin buys 12 kilograms of birdseed.
The birds eat 3 kilograms of birdseed.
How much birdseed is left? ________ kilograms

Spiral Review

Draw the clock hands to show the end time.

4.

2 hours later:

Name ____________________

Liter

Choose the better estimate.

1. 

about 2 liters (circled)

about 200 liters

about 1 liter

about 10 liters

2. 

about 4 liters

about 400 liters

about 4 liters

about 400 liters

Problem Solving

Draw a picture. Solve.

3. Ellie buys 4 liters of juice for her birthday party. Each liter fills 4 glasses. How many glasses can be filled?

________ glasses

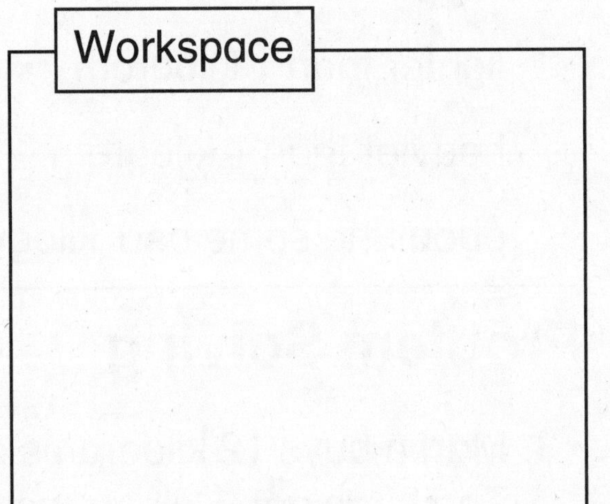

Spiral Review

4. Find the perimeter of the figure. Use an inch ruler.

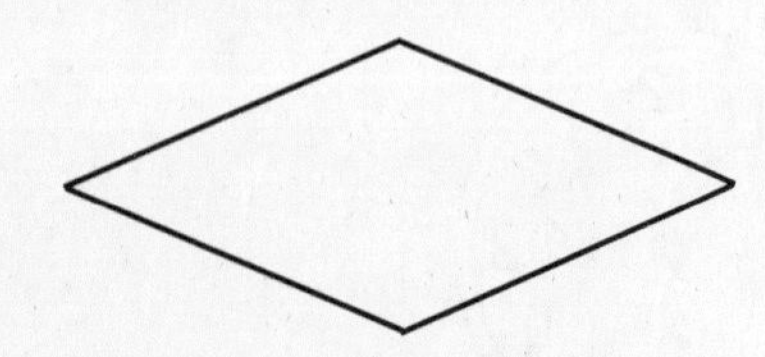

___ + ___ + ___ + ___ = ___ inches

Name________________________________

Temperature

Write each temperature.

1.

___30___ °F

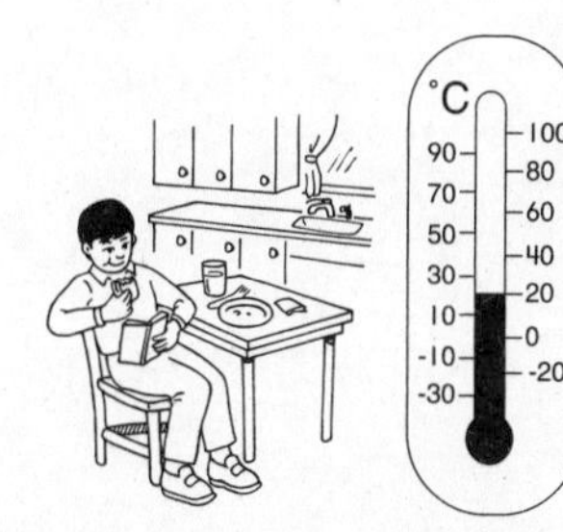

________ °C

2.

________ °C

________ °F

Solve.

3. It is 62°F at 8:00 in the morning. The radio says it will be 20 degrees warmer by 3:00 in the afternoon. What temperature will it be at 3:00? ________ °F

Spiral Review

Use the data. Answer the question.

The following numbers show how many trees the second graders planted in the past five weeks.

4, 7, 6, 12, 7

4. What is the range of the numbers? ________

Name ______________________________

Measurement Tools

Match to show the tool you can use to measure.

1. 

How cold is it?

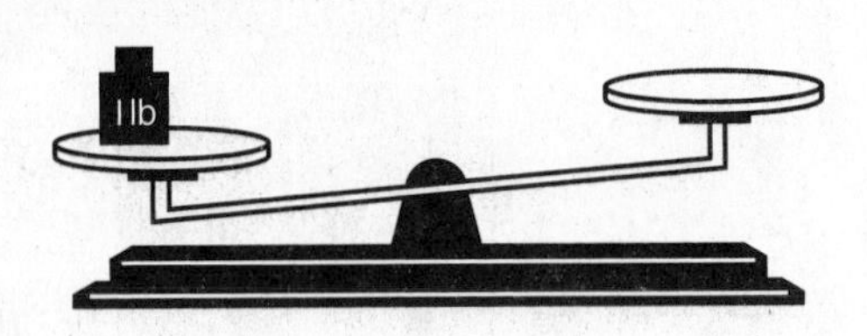

2. 

How heavy is it?

3.

How long is it?

Solve.

4. It is 5° C when Anissa wakes up. After school, the temperature is 20° C. How many degrees warmer is it after school than it was in the morning? ________ ° C

Spiral Review

Color to show the number of square units.

5. 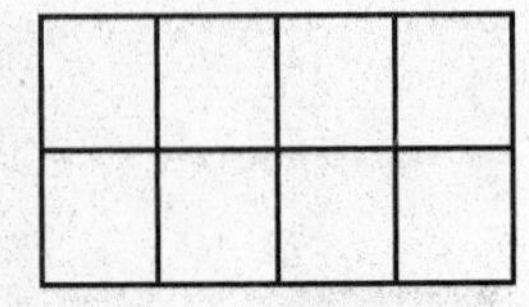

2 square units

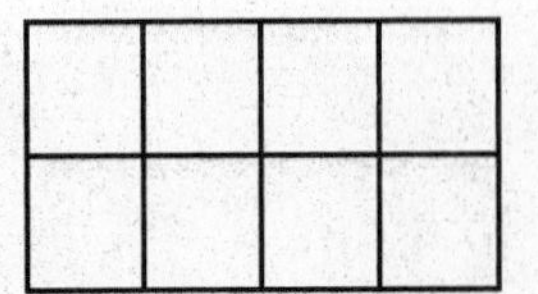

5 square units

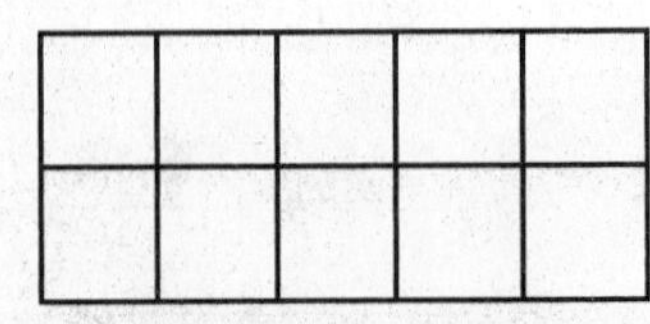

7 square units

Name ____________________

Solid Figures

Circle the solid figure that answers each question.

1. Which has 6 faces?

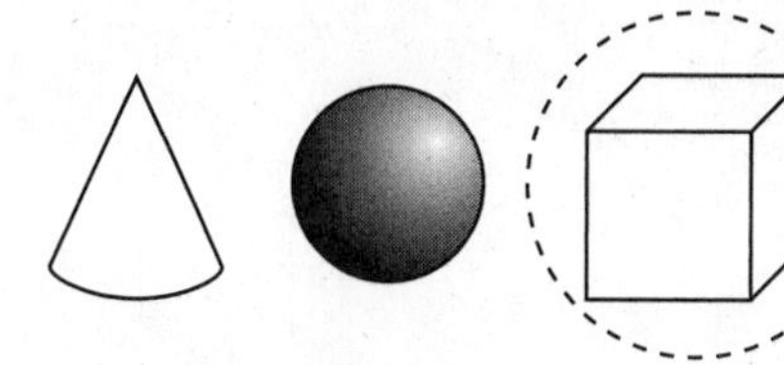

2. Which has 8 vertices?

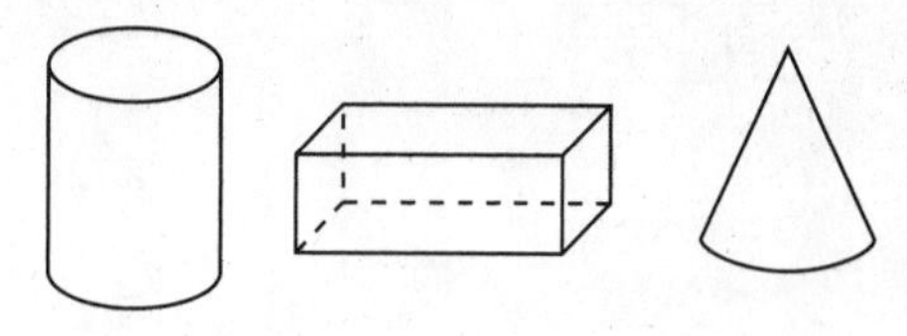

3. Which has 8 edges?

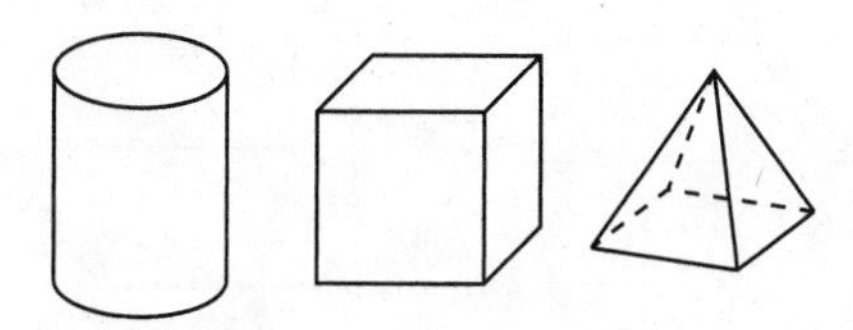

4. Which has 5 faces?

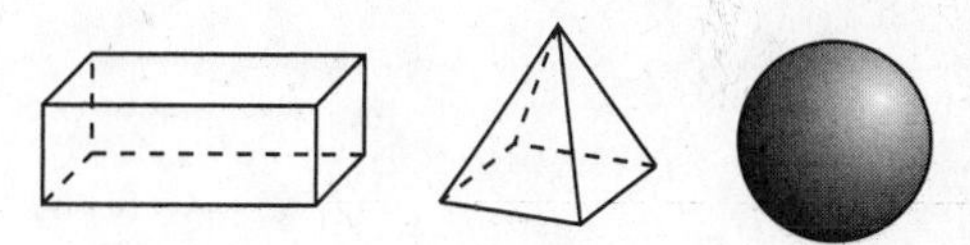

Problem Solving

5. I have 8 vertices and all my faces are the same size. Which solid figure am I?

____

6. I have 5 vertices and 5 faces. Which solid figure am I?

____

Spiral Review

Add or subtract.

7.

5	12	8	14	6	9	17
+ 6	− 9	+ 7	− 5	+ 7	+ 7	− 9

Name______________________________

Solid and Plane Figures

Circle each solid figure you can trace to make the shapes.

1.

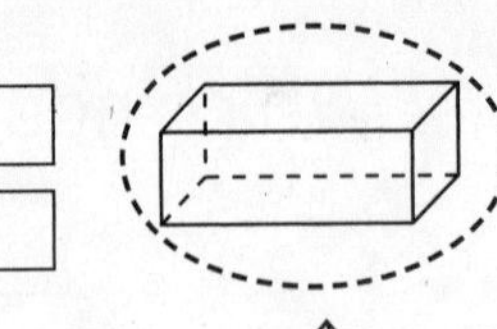

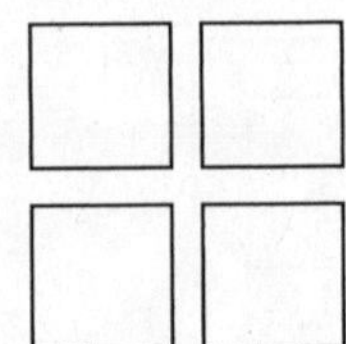

2.

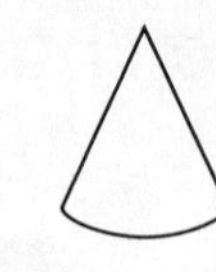

3.

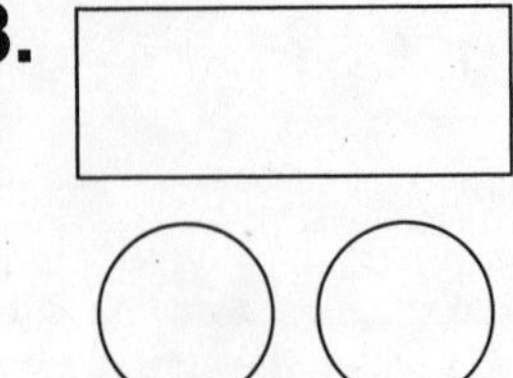

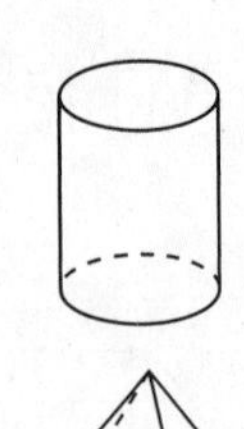

4.

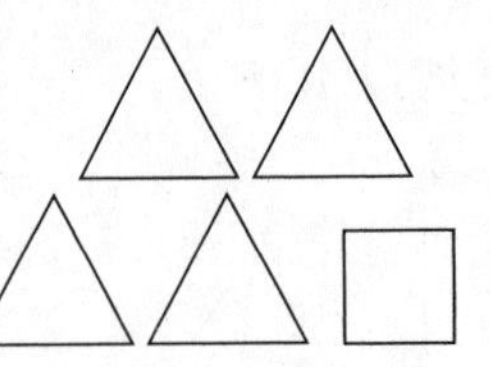

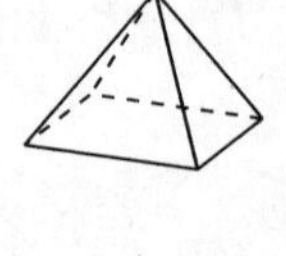

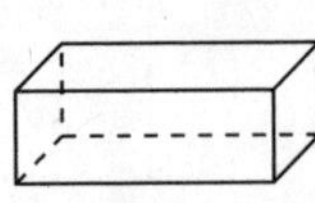

Problem Solving

5. Weng and Selena have 40 blocks. Weng builds this house. Are there enough blocks for Selena to build the same house?

______

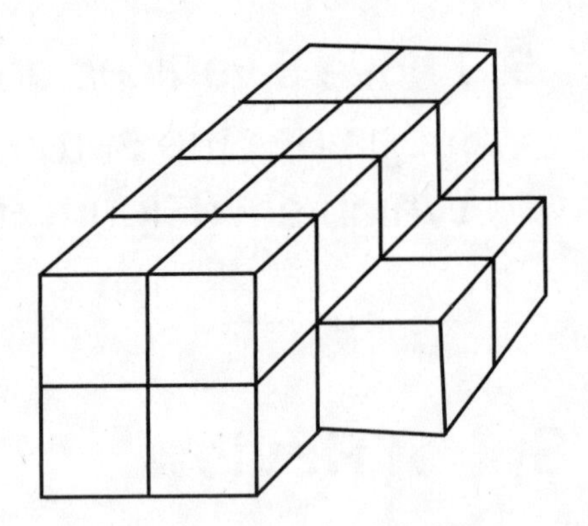

Spiral Review

Write the sentence in a different order. Add.

6. 44 + 48 = ______ 21 + 55 = ______ 28 + 13 = ______

____ + ____ = ____ ____ + ____ = ____ ____ + ____ = ____

Name ____

More Plane Figures

Color the figures named. Then tell how many sides and angles each has.

1. parallelogram

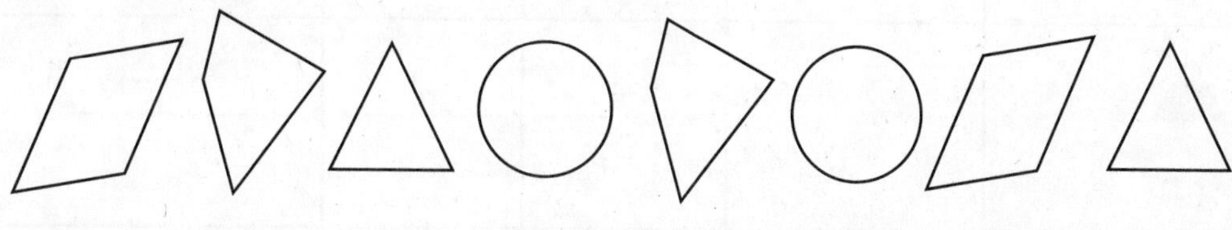

<u>4</u> sides <u>4</u> angles

2. quadrilateral

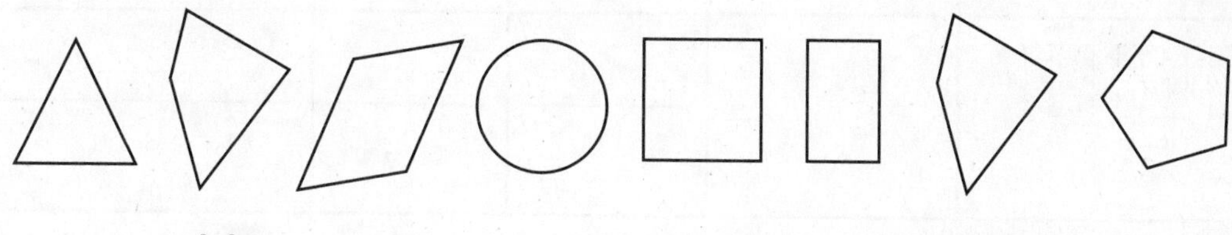

______ sides ______ angles

3. pentagon

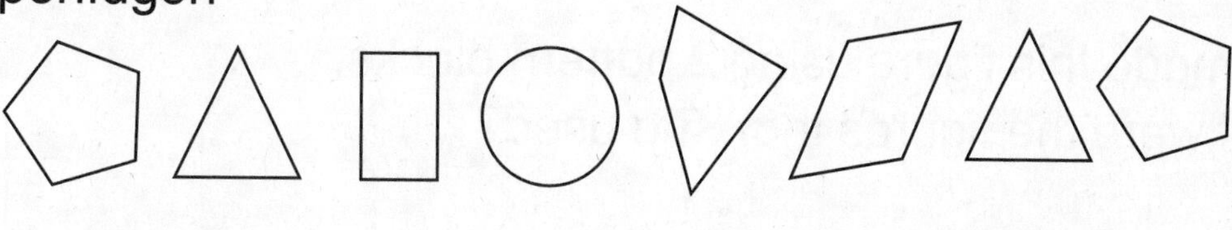

______ sides ______ angles

Spiral Review

Look at the calendar and write the answers.

December 2002						
Sunday	Monday	Tuesday	Wednesday	Thursday	Friday	Saturday
1	2	3	4	5	6	7
8	9	10	11	12	13	14
15	16	17	18	19	20	21
22	23	24	25	26	27	28
29	30	31				

4. On what day of the week does December begin? __________

5. How many days are there between December 15 and 25? __________

Name ___________________________________

Make Figures

Use pattern blocks to make new figures. Complete the chart.

Use these blocks.	Make a new figure.	How many sides?	How many corners?	Name of new figure
1. ▲▲		4	4	parallelogram
2. (2 trapezoids)		____	____	____________
3. ■■■		____	____	____________

Problem Solving

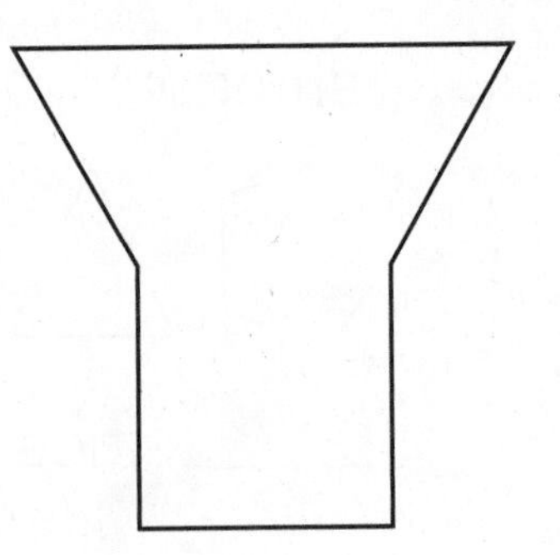

4. Kira made this figure using 2 pattern blocks. What were the figures that Kira used?

____

Draw a line to show how Kira put the figures together.

Spiral Review

Subtract.

5.
$$\begin{array}{r} 65 \\ -\ 9 \\ \hline \end{array} \quad \begin{array}{r} 59 \\ -\ 27 \\ \hline \end{array} \quad \begin{array}{r} 73 \\ -\ 47 \\ \hline \end{array} \quad \begin{array}{r} 66 \\ -\ 38 \\ \hline \end{array} \quad \begin{array}{r} 84 \\ -\ 31 \\ \hline \end{array} \quad \begin{array}{r} 75 \\ -\ 57 \\ \hline \end{array}$$

Name______________________________

Problem Solving: Reading for Math

Make Decisions

Jessie builds a house. She uses different figures.

Build a house. Use different figures. Then answer the questions.

1. Which figures did you use to make the front of the house? Why?

   ____

2. Which figures did you use to make the roof? Why?

   ____

3. What does a side view of the house look like? Draw the house.

   ____

Workspace

Spiral Review

Draw a picture. Solve.

4. 24 people are on the bus.
   15 people get off at 34th Street.
   How many people are left on the bus? ________ people

5. 13 people are in line for the movie.
   8 more people get in line.
   How many people are in line altogether? ________ people

Name ___________________________________

Problem Solving: Strategy
Act It Out

Use pattern blocks. Tell how many triangles and squares are used to make each figure.

1.

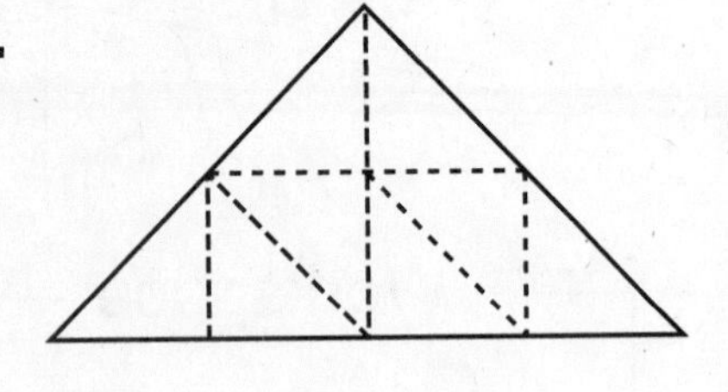

<u>8</u> triangles

2.

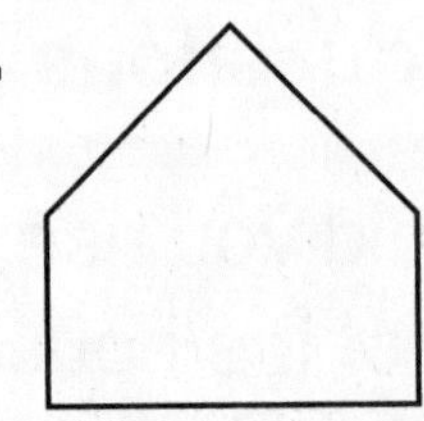

_____ triangles _____ squares

3.

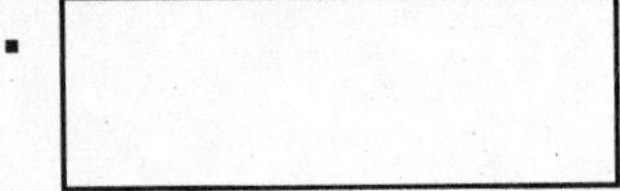

_____ triangles

4.

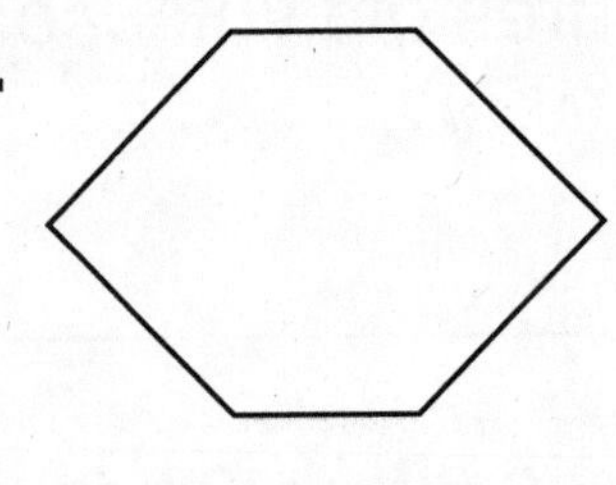

_____ triangles _____ squares

Spiral Review

Circle *add* or *subtract*. Solve.

5. It snows on 13 days in February. February has 28 days. How many days in February get no snow?

add subtract

____

_____ days

6. There are 8 days of rain in March. There are 14 days of rain in April. How many days does it rain in March and April in all?

add subtract

____

_____ days

Name______________________________

Congruent Figures

Draw a figure that is congruent.

1.

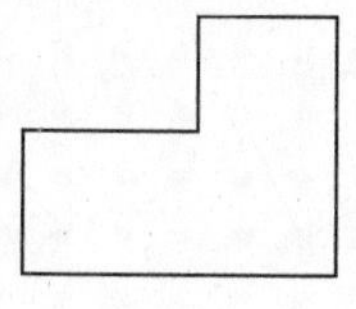

2.

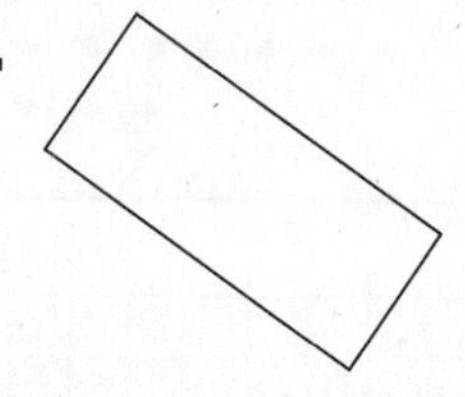

Color the two congruent figures.

3.

 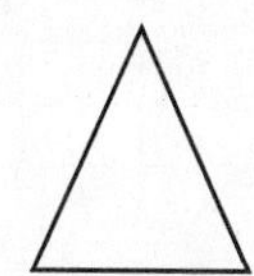 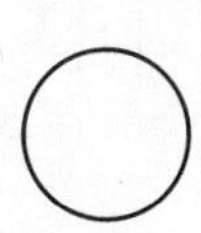  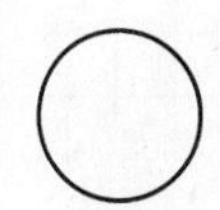

4.

 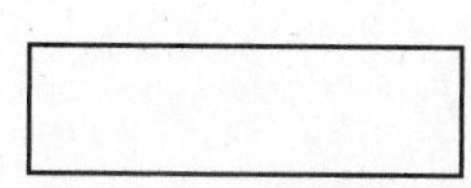 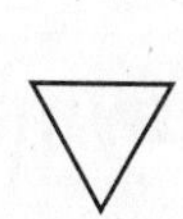 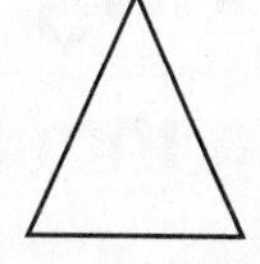 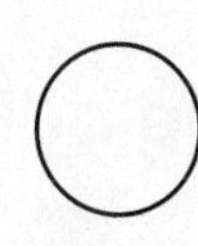

5.

 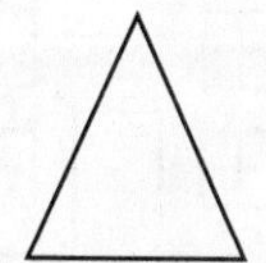 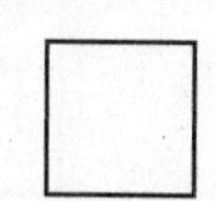 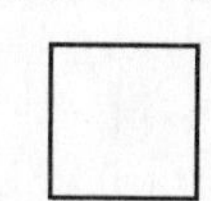 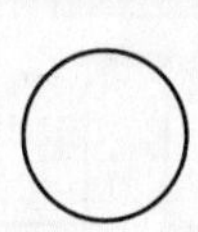

Spiral Review

Complete each sentence. Use >, <, or =.

6. 64 ◯ 46   23 ◯ 32   51 ◯ 5 tens + 1 one

7. 75 ◯ 70 + 5   15 ◯ 36   88 ◯ 22

Name ___________________________________

Symmetry

Draw a matching part for each figure.

1. 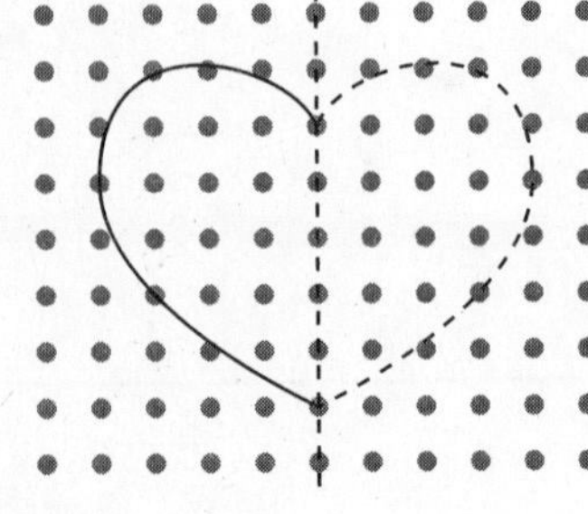

2. 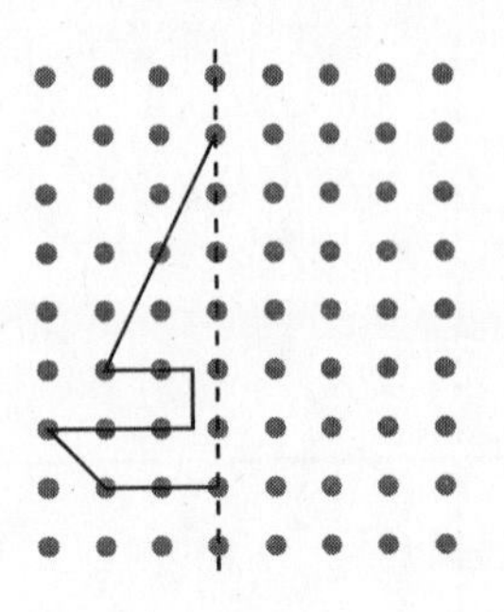

3. 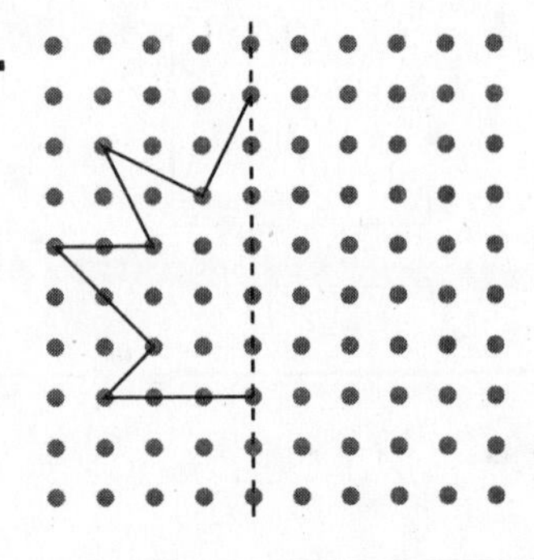

4. 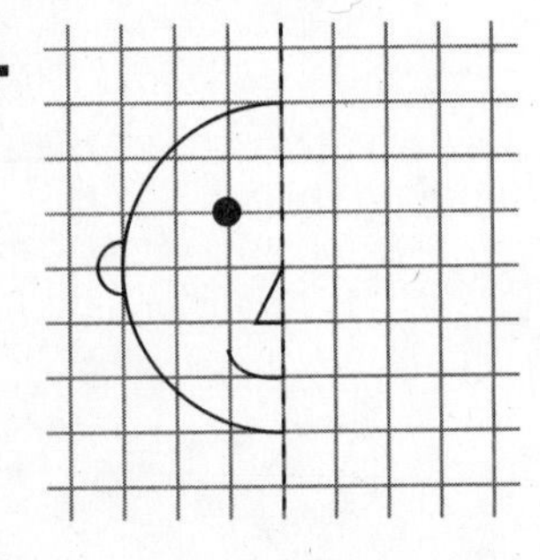

5. 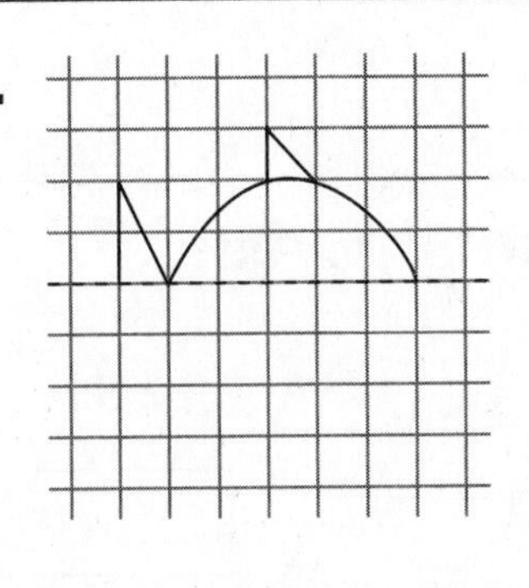

6. 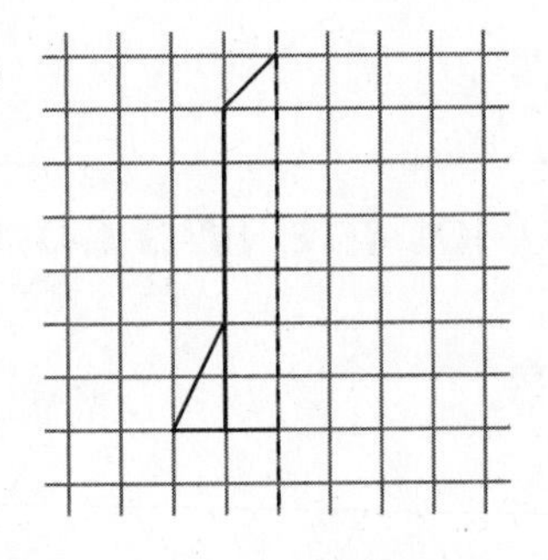

Problem Solving

Circle the building that has symmetry.

7. 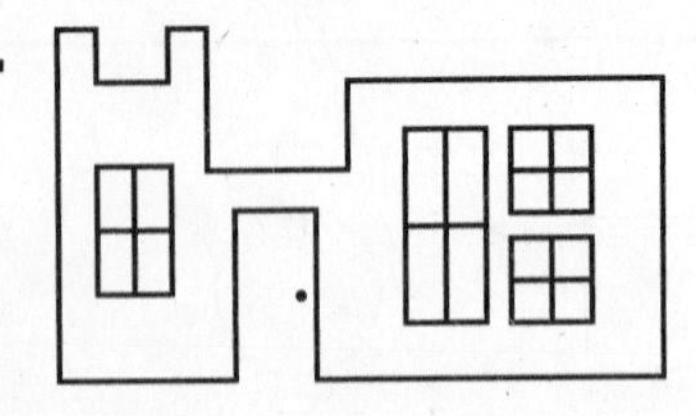 

Spiral Review

Circle the tool you use to measure.

8. How long is a pencil?

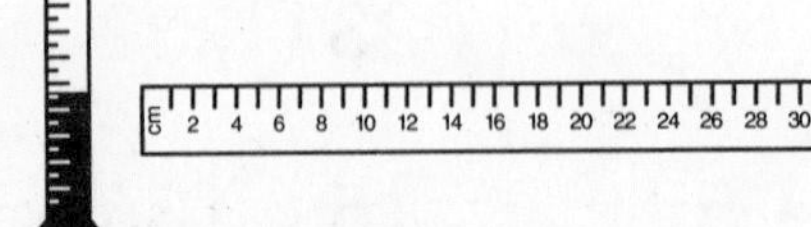

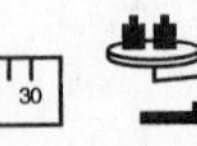

Name ____________________________________

Halves, Fourths, and Eighths

Color one part of each figure.
Then write the fraction for each part.

1. 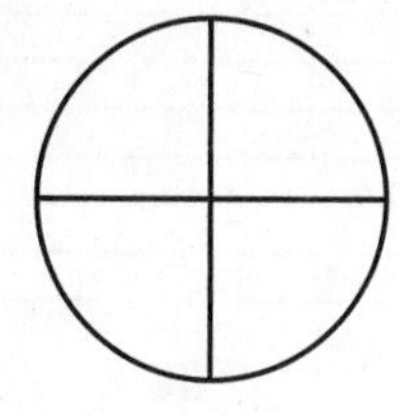 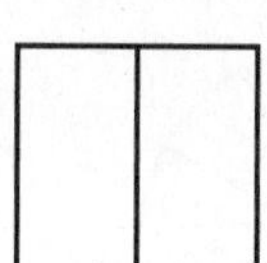 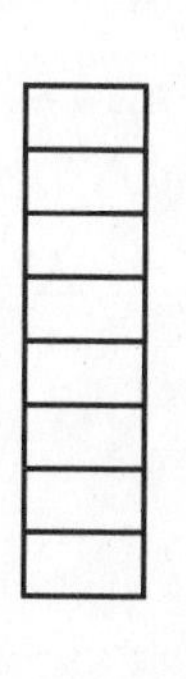

$\frac{1}{4}$ ______ ______

Answer each question. Write yes or no.

2. Is this $\frac{1}{2}$? ______ 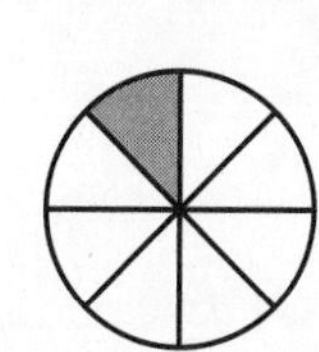 Is this $\frac{1}{8}$? ______

Spiral Review

Use the diagram to answer the questions.

3. How many children like books about people?

______ children

4. How many children like books about animals and books about people? ______ children

Name ____________________

Thirds, Sixths, and Twelfths

Color one part of each figure.
Then write the fraction for the part.

1.

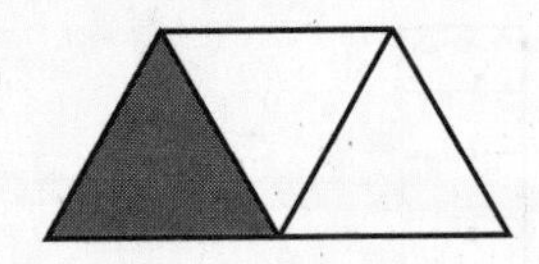

$\frac{1}{3}$ ______ ______

Write the word and the fraction for each part.

2. Each part is one __________, or ______.

3. Each part is one __________, or ______.

Spiral Review

Use data from the table to answer the question.

4. How many more cats were at the pet show than fish?

__________ cats

Second Grade Pet Show

Cats	𝍸 𝍸 II
Dogs	𝍸 III
Fish	𝍸
Hamsters	III

Name________________________________

More Fractions

Remember, fractions can name more than one equal part.

Write the fraction for the part that is shaded.

1. 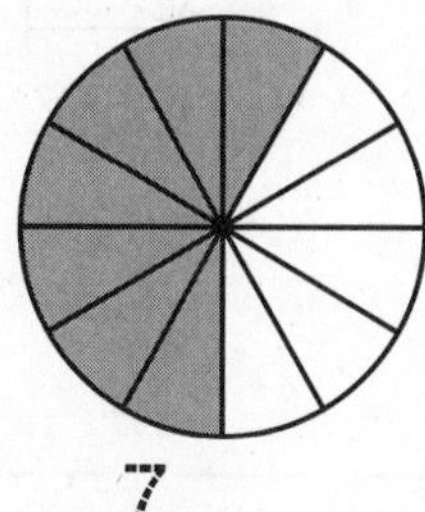

$\frac{7}{12}$

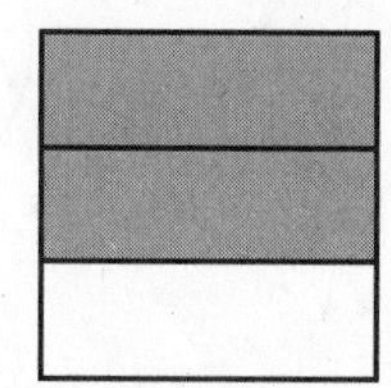

______

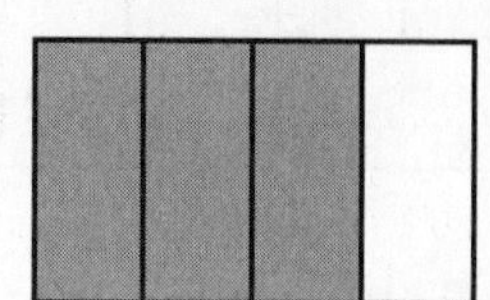

______

Problem Solving

Look at the picture. Solve.

2. The fish tank is almost full. What fraction of the tank has water in it?

______

Spiral Review

Color to show the number of square units.

3. 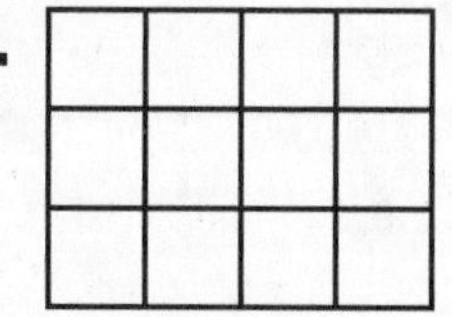

8 square units

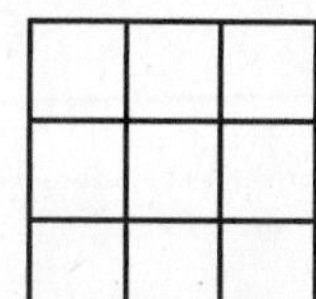

5 square units

Name ________________________________

Compare Fractions

Circle the fraction that is greater.

1. 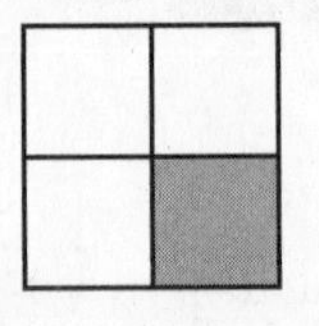 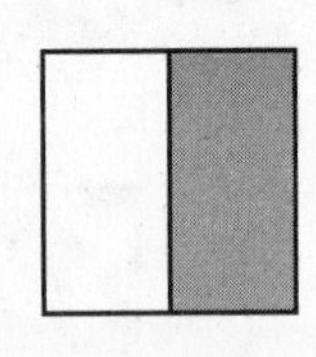

$\frac{1}{4}$ $\frac{1}{2}$

2. 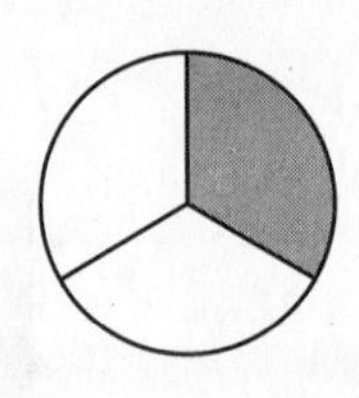 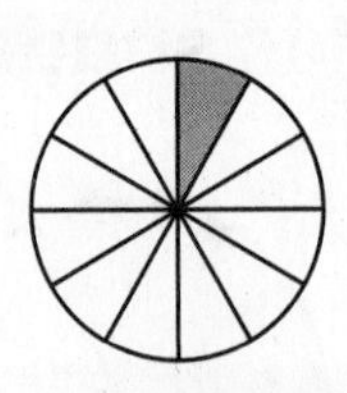

$\frac{1}{3}$ $\frac{1}{12}$

Compare the fractions. Use >, <, or =.

3. $\frac{1}{3} \bigcirc \frac{1}{2}$ $\frac{1}{3} \bigcirc \frac{1}{4}$ $\frac{1}{6} \bigcirc \frac{1}{12}$

Solve.

4. Ella has $\frac{1}{6}$ of a pie.

George has $\frac{1}{8}$ of a pie.

Who has the bigger piece of pie?

____

Spiral Review

Write each temperature.

5. 

____

____

Name

Problem Solving: Reading for Math

Making Predictions

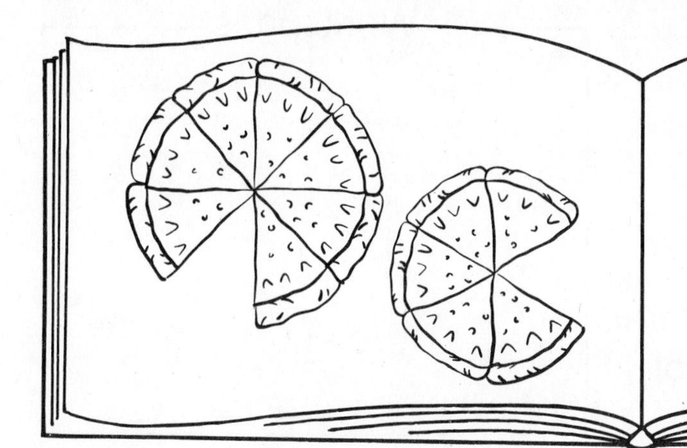

Jenna and Max ate peach pie at a picnic.

Max ate $\frac{1}{6}$ of a pie.

Jenna ate $\frac{1}{8}$ of a pie.

Predict which pie Max ate.

1. How much of the large pie is left? ______
2. How much of the small pie is left? ______
3. Which pie do you think Max ate? ______
4. Is it possible that Jenna ate more than Max? Explain.

______

______

Spiral Review

Count to find the total amount.

Do you have enough money to buy the pens? Circle *yes* or *no*.

5. 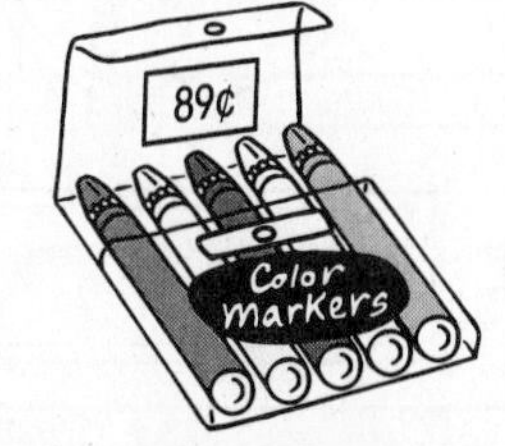

yes no

Name ______________________________

11·6 **Problem Solving: Strategy**

Draw a Picture

Draw a picture. Solve.

Workspace

1. Jen and Seth paint a big square on the playground. They divide the square into 4 equal parts. They paint 2 of the 4 parts blue. What fraction of the square is blue? $\frac{2}{4}$

2. A pizza is cut into 12 pieces. 5 of the pieces have peppers on them. What fraction of the pizza has peppers on it? ______

3. Nick bakes a pan of brownies. He divides the pan into 12 parts. He puts frosting on 8 parts. What fraction of the brownie pan has frosting on it? ______

Spiral Review

Write the time for each clock.

4.

Name ____________________

Fractions of a Group

Color to show each fraction.

1. $\frac{1}{3}$ ◯ ◯ ◯

2. $\frac{3}{4}$ ◯ ◯ ◯ ◯

Look at the picture. Find the fraction.

3. What fraction of the apples are shaded?

___ → number of shaded apples

___ → total number of apples

4. What fraction of the cherries are shaded?

___ → number of shaded cherries

___ → total number of cherries

Look at the picture. Answer the question.

5. Kyra has 6 cookies. 2 are oatmeal cookies. 4 are chocolate cookies. What fraction of the cookies are oatmeal?

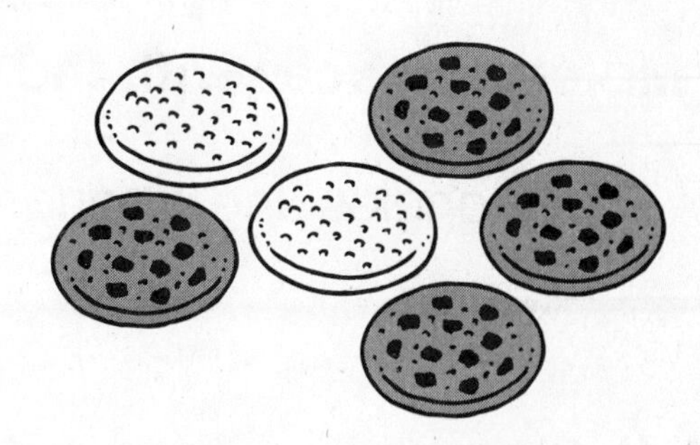

____

Spiral Review

Complete each sentence. Use >, <, or =.

6. 78 ◯ 31   34 ◯ 43   26 ◯ 2 tens + 6 ones

Name ______________________________

11·8 More Fractions of a Group

Color to show the fraction.

1. $\frac{1}{2}$ of the circles are red.

$\frac{3}{4}$ of the circles are orange.

$\frac{1}{3}$ of the circles are yellow.

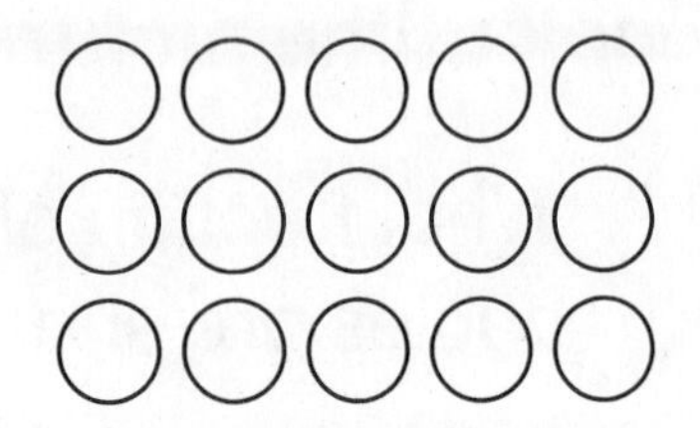

Answer the question.

2. Zoe has 2 red apples and 6 yellow apples.
How many apples does she have altogether? ______ apples

What fraction of the apples are yellow? __________

Spiral Review

Match the faces with the solid figures.

3. 4 triangles and 1 square

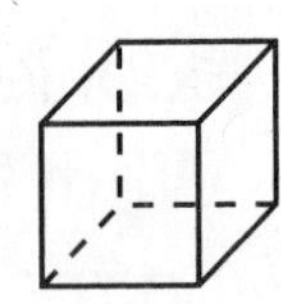

4. 2 circles

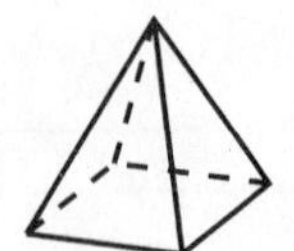

5. 6 squares

Name ______________________________________

Most Likely and Least Likely Outcomes

Look at the items. Circle the one you are most likely to pick. Circle the one you are least likely to pick.

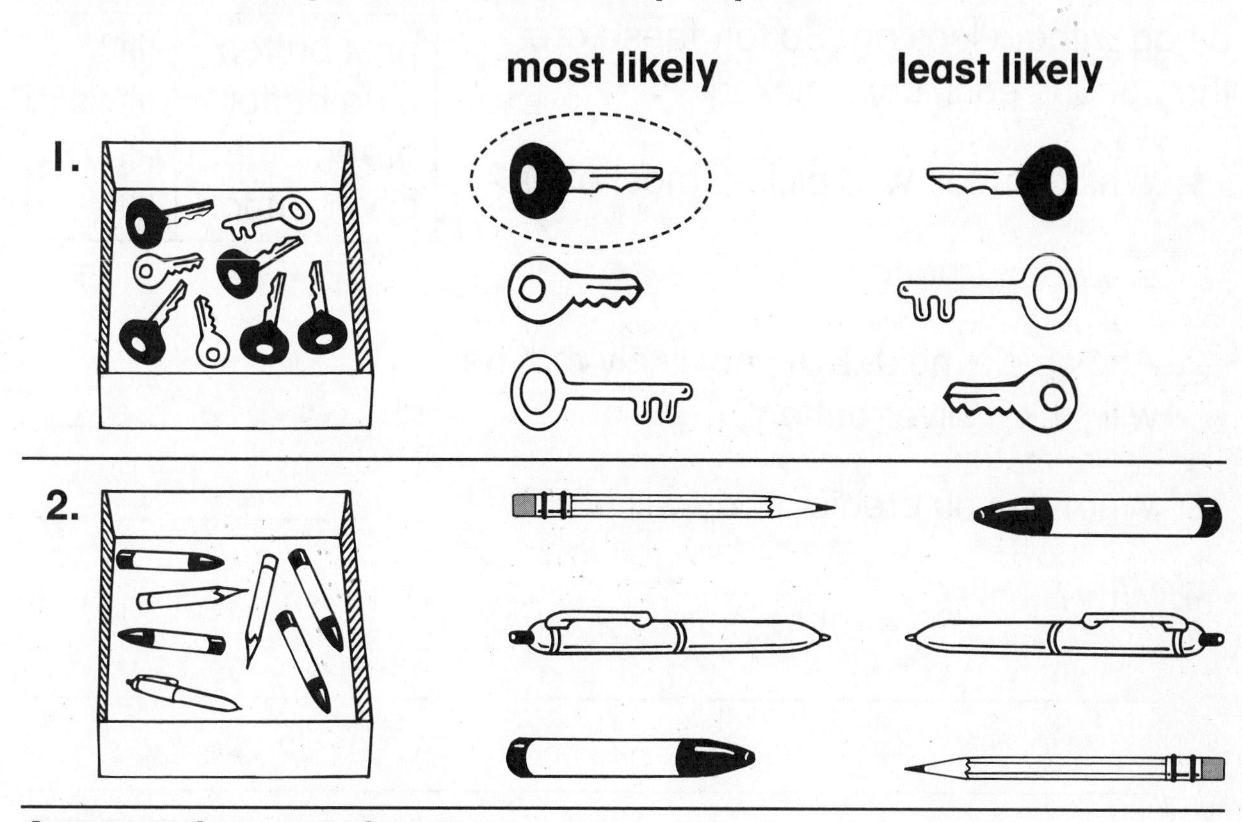

Answer the questions.

3. Edna has 8 candies in a bag. 5 candies are silver, 2 are green, and 1 is blue. Edna chooses a candy without looking. Which color is Edna most likely to pick? ______________

Which color is Edna least likely to pick? ______________

Spiral Review

Write the next three even numbers.

4. 24 ______ ______ ______ 48 ______ ______ ______

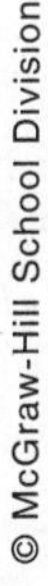

Name ______________________________

Make Predictions

Answer each question.

Each second grader picked a button from a bag without looking. So far, these are the buttons each child picked.

Buttons

pink button	||||/	
blue button	||||/ ||||/ ||
green button	||||/
silver button	||||

1. Which button was picked most often?

   ________ button

2. Andy picks next. Is it very likely that he will pick a silver button? ____________

3. What do you predict Andy will pick? ____________

4. Why?

   ____

   ____

Spiral Review

Write the number that comes between.

5.

59	14	37	20
____	____	____	____
61	16	39	22

Name ______________________

Hundreds

Write how many.

1. 6 groups of hundreds

__6__ hundreds =

__600__ in all

4 groups of hundreds

______ hundreds =

______ in all

2. 8 groups of hundreds

______ hundreds =

______ in all

5 groups of hundreds

______ hundreds =

______ in all

Problem Solving

4. Which group has more?

____

How do you know?

____

____

____

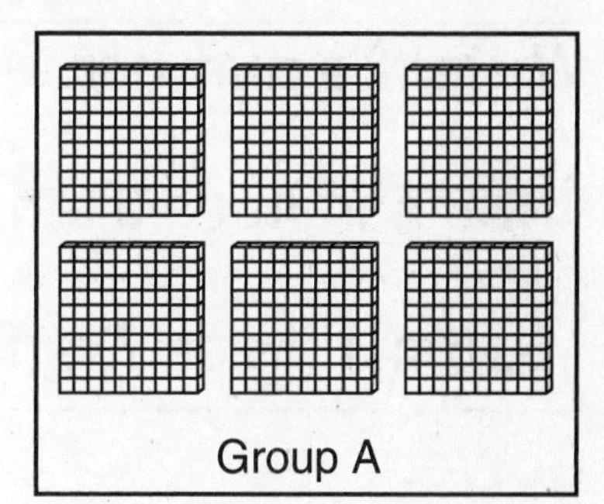
Group A

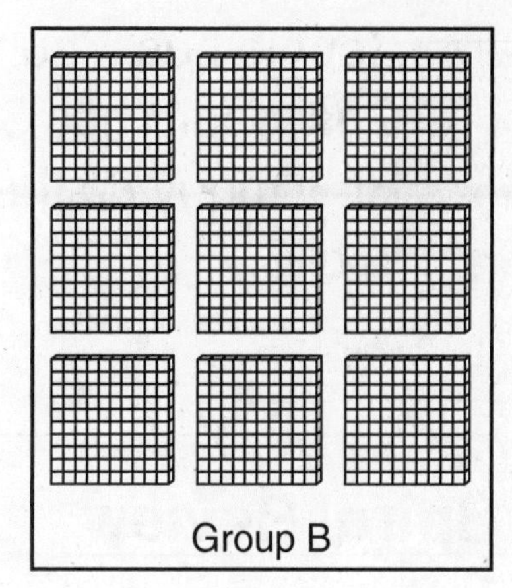
Group B

Spiral Review

Add.

5. 2 + 6 = ______ 2 + 3 = ______ 4 + 3 = ______

20 + 60 = ____ 20 + 30 = ____ 40 + 30 = ____

Name ____________________

Hundreds, Tens, and Ones

Write how many hundreds, tens, and ones.

1. 357 __3__ hundreds

__5__ tens __7__ ones

hundreds	tens	ones
3	5	7

2. 424 ______ hundreds

______ tens ______ ones

hundreds	tens	ones

3. 740 ______ hundreds

______ tens ______ ones

hundreds	tens	ones

Write the number.

4. 8 hundreds 0 tens 3 ones ________

5. 5 hundreds 7 tens 2 ones ________

Solve.

6. Sophie has 5 hundreds. Cassandra has 3 hundreds. How many do they have in all?

____

7. Nate has 9 hundreds. Joe has 4 hundreds. How many more hundreds does Nate have?

____

Spiral Review

8. Color to show the fraction one fourth. Write the fraction. ______

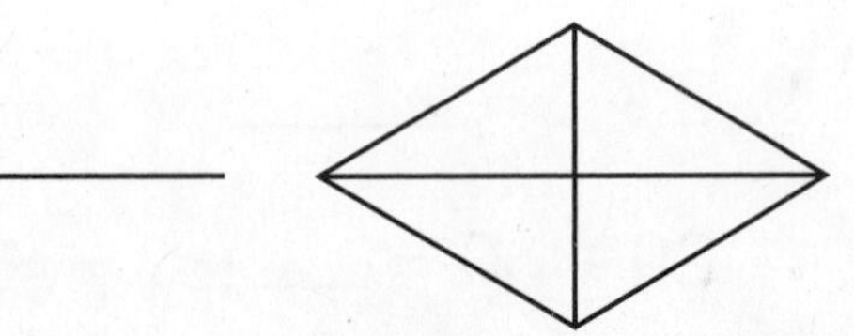

Name ______________________________

More Hundreds, Tens, and Ones

Circle the value of each underlined digit.

1.	<u>4</u>59	4 hundreds	5 tens	9 ones
2.	5<u>0</u>8	5 hundreds	0 tens	8 ones
3.	7<u>2</u>4	7 hundreds	2 tens	4 ones
4.	<u>5</u>96	5 hundreds	9 tens	6 ones
5.	32<u>4</u>	3 hundreds	2 tens	4 ones
6.	5<u>7</u>0	5 hundreds	7 tens	0 ones
7.	28<u>9</u>	2 hundreds	8 tens	9 ones

Solve.

8. What is the greatest 3-digit number you can write using the digits 2, 8, and 6? ________

9. How many hundreds, tens, and ones are in 461?

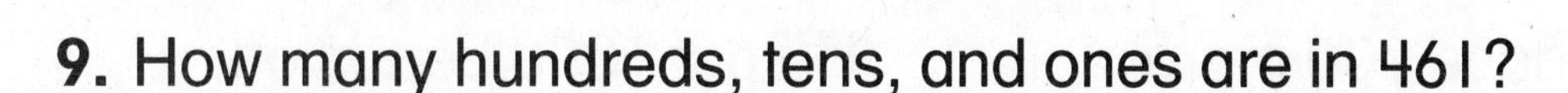

Spiral Review

10. Use a centimeter ruler. Measure each side. Write about how many centimeters around the shape is.

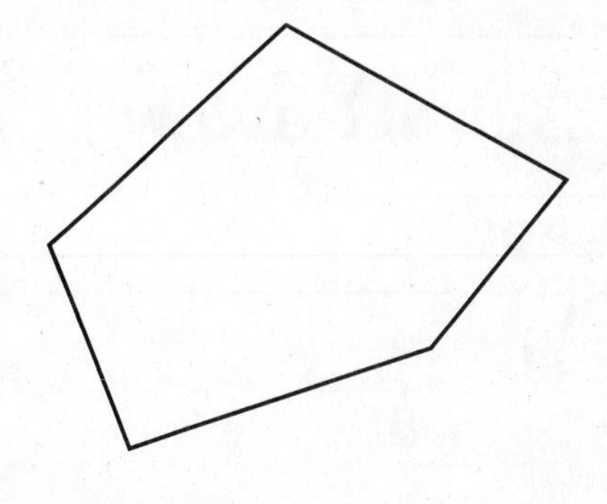

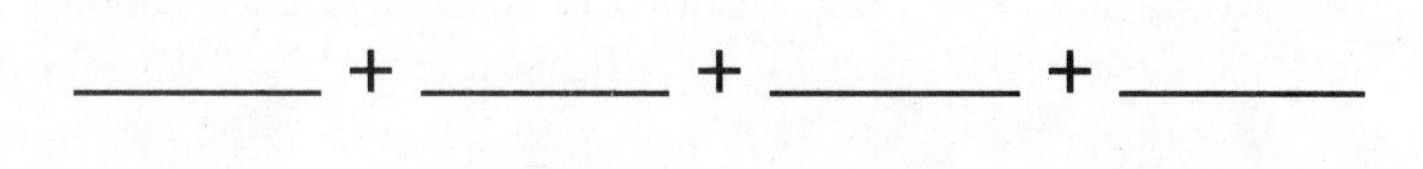

\+ ______ = ______ centimeters

Name ___________________________________

Counting On or Back by Hundreds

Count on or back by hundreds. Write each number.

1. One hundred more than 841 is ___941___.
2. One hundred less than 602 is _________.
3. One hundred more than 263 is _________.
4. One hundred less than 316 is _________.
5. Two hundred more than 555 is _________.
6. Three hundred more than 128 is _________.
7. Three hundred less than 754 is _________.
8. Four hundred more than 250 is _________.

Problem Solving

9. Holly has 139 buttons in a jar. She adds 200 more buttons to the jar. How many buttons does Holly have in all?

   _________ buttons

10. Keith has 567 beads. He puts 300 beads on a string. How many beads does Keith have left?

    _________ beads

Spiral Review

Add.

11. $\begin{array}{r} 8 \\ +8 \\ \hline \end{array}$ $\begin{array}{r} 7 \\ +4 \\ \hline \end{array}$ $\begin{array}{r} 4¢ \\ +9¢ \\ \hline \end{array}$ $\begin{array}{r} 6 \\ +9 \\ \hline \end{array}$ $\begin{array}{r} 8 \\ +5 \\ \hline \end{array}$ $\begin{array}{r} 7 \\ +8 \\ \hline \end{array}$

Name____________________

Numbers to 1,000

Write each number. Write each word name.

hundreds	tens	ones
4	1	9

1. ________

____

hundreds	tens	ones
3	5	3

2. ________

____

hundreds	tens	ones
5	2	0

3. ________

____

Problem Solving

4. Look at the picture.
How many beads are there?
Write the number. ________

Write the word. ________________

100

200

200

300

Spiral Review

Subtract. Do you need to regroup? Circle *yes* or *no*. Solve.

5. 29 – 7 yes no ________

46 – 9 yes no ________

Name ______________________________

Expanded Form

Write each number in expanded form.

1. 566   500 + 60 + 6
2. 830   ______ + ______ + ______
3. 271   ______ + ______ + ______
4. 304   ______ + ______ + ______
5. 989   ______ + ______ + ______
6. 742   ______ + ______ + ______

Solve.

7. Charlie glues 246 buttons to his puppet. Write this number in expanded form.

____

8. Julie has 7 beads. Mona has 20 beads. Zack has 100 beads. How many beads do they have in all?

____

Spiral Review

9. Circle the clock that shows the same time.

8:30

Name ______________________________

Problem Solving: Reading for Math

Problems and Solutions

Solve.

Maggie is threading buttons.
She chooses 6 black buttons, 3 gray buttons, and 3 white buttons.
What pattern does Maggie make?

1. How many black buttons are on the string? __________

2. What pattern is Maggie making with the buttons?

____

3. What color is the last button Maggie will use? __________

Spiral Review

Circle groups of 2. Write *even* or *odd*.

4. 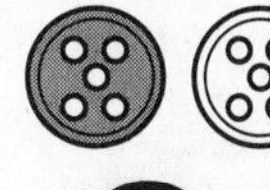  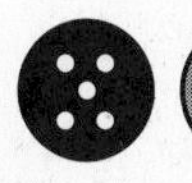 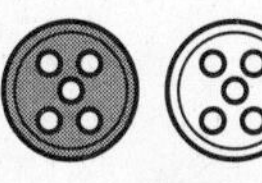  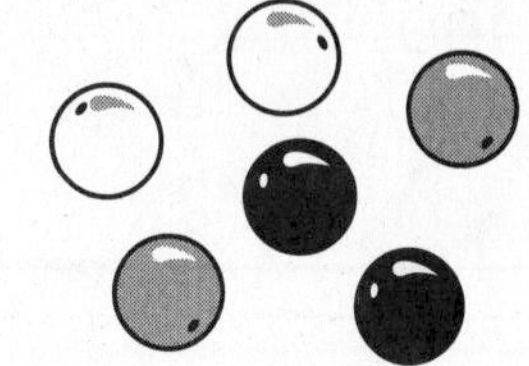 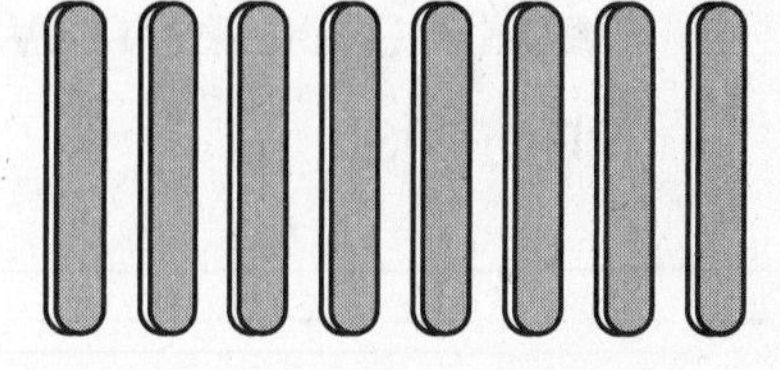

__________ __________ __________

__________ __________ __________

Name

Problem Solving: Strategy

Find a Pattern

Continue each pattern. Write or draw what would most likely come next.

1. 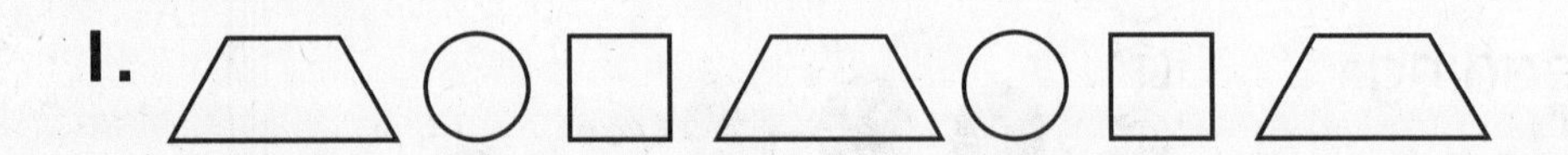 circle

2.        ______

3.  ______

4.  ______

Solve.

5. Valerie has 10 charms.
5 charms are gold and
5 charms are silver.
Draw a picture.
Make a pattern with the charms.

Spiral Review

6. 3 + 6 + 6 = ______ 5 + 5 + 7 = ______ 8 + 2 + 5 = ______

Name ______________________________

Compare Numbers

Compare. Write >, <, or =.

1. 864 __<__ 865	812 ____ 822	678 ____ 678	410 ____ 310
2. 578 ____ 516	483 ____ 485	560 ____ 555	960 ____ 860
3. 502 ____ 502	122 ____ 120	844 ____ 844	280 ____ 279
4. 599 ____ 600	325 ____ 323	167 ____ 761	862 ____ 873
5. 467 ____ 458	912 ____ 932	343 ____ 343	760 ____ 587

Problem Solving

6. I am greater than 4 hundreds 3 tens and 3 ones.I am less than 4 hundreds 3 tens and 5 ones. What number am I?

____

7. Nancy is waiting for a bus. The number of the bus is greater than 1 hundred 9 tens and 2 ones. The number is less than 1 hundred 9 tens and 4 ones. What number is the bus?

____

Spiral Review

Add. Look for patterns.

8. 30 + 10 = ________ 20 + 15 = ________

30 + 20 = ________ 20 + 20 = ________

30 + 30 = ________ 20 + 25 = ________

30 + 40 = ________ 20 + 30 = ________

Name ______________________________

Order Numbers

Write the number that comes just after.

1. 764 765 207 ______ 652 ______ 284 ______

2. 835 ______ 914 ______ 101 ______ 342 ______

Write the number that comes just before.

3. ______ 197 ______ 285 ______ 987 ______ 659

4. ______ 571 ______ 455 ______ 621 ______ 377

Write the number that comes between.

5. 288 ______ 290 405 ______ 407 961 ______ 963

6. 755 ______ 757 682 ______ 684 510 ______ 512

Problem Solving

Circle your answer.

7. Which number is closest to 500? 498 510 501

8. Which number is closest to 800? 750 825 850

Spiral Review

Write the number in three different ways.

9. 

______ tens ______ ones

______ + ______

______

Name ____________________

Number Patterns

The numbers go up by ones, tens, or hundreds. Write the missing numbers in each pattern. Then circle the pattern.

	Numbers	Pattern: Count By
1.	810, 820, 830, 840, 850, 860	ones (tens) hundreds
2.	400, 500, 600, ____, 800, ____	ones tens hundreds
3.	345, 346, ____, 348, 349, ____	ones tens hundreds
4.	425, 525, 625, ____, ____, 925	ones tens hundreds
5.	523, 524, ____, 526, 527, ____	ones tens hundreds

Problem Solving

6. Pedro counts pennies in groups of ten.
How many pennies does Pedro have if he has 6 groups?

Number of groups	1	2	3	4	5	6
Number of pennies	10	20	30	____	____	____

Spiral Review

Complete the number sentence.

7. ____ + 9 = 12 11 = ____ + 4 5 + ____ = 10

Name ______________________________

Add Hundreds

Count on by hundreds to add.

1. 200 + 600 = 800   100 + 300 = ______   300 + 400 = ______

2. 500 + 100 = ______   600 + 300 = ______   200 + 700 = ______

3.

200	500	300	400	700	600
+ 200	+ 400	+ 300	+ 500	+ 200	+ 100

4.

400	100	400	300	500	600
+ 200	+ 400	+ 300	+ 600	+ 200	+ 200

Problem Solving

5. The school kitchen collects food scraps to compost. In one month, the kitchen collected 300 pounds of scraps. The next month they collected 400 pounds of scraps. How many pounds of scraps were collected in all?

   ______ pounds

6. The school sponsors a car wash. The first graders wash 100 cars. The third graders wash 400 cars. How many cars are washed in all?

   ______ cars

Spiral Review

Write how many hundreds, tens, and ones.

7. 452 ______ hundreds ______ tens ______ ones

Name ______________________________

3-Digit Addition Without Regrouping

Find each sum.

1.

	hundreds	tens	ones
	5	1	4
+	4	5	4
	9	6	8

	hundreds	tens	ones
	1	7	3
+		2	6

	hundreds	tens	ones
	2	8	3
+	1	0	6

2. $538 + 241$ $614 + 173$ $936 + 21$ $318 + 600$ $211 + 333$

3. $423 + 60$ $256 + 132$ $333 + 52$ $481 + 113$ $325 + 100$

Problem Solving

4. Mary collects 145 cans for recycling. Sarah collects 234 cans. How many cans are collected in all?

______ cans

5. What is the same about the numbers 529 and 259? What is different?

____

____

____

____

Spiral Review

Count forward. Write the missing numbers.

6. 4, ______, 6, ______, 8

Name

3-Digit Addition

Add. Use hundreds, tens, and ones models.

1.

	hundreds	tens	ones
	1		
	2	6	4
+	4	5	4
	7	1	8

	hundreds	tens	ones
	3	7	5
+		1	6

	hundreds	tens	ones
	2	8	3
+	3	4	6

2.

	hundreds	tens	ones
		1	
	2	6	7
+	3	2	4

	hundreds	tens	ones
	5	4	6
+	1	2	6

	hundreds	tens	ones
	4	4	3
+	1	4	9

Problem Solving

3. Draw a picture.
Show the sum of 154 and 221. ______

Spiral Review

Add.

4.

200	300	300	100	700	600
+ 200	+ 400	+ 300	+ 500	+ 200	+ 300

Name___________________________

More 3-Digit Addition

Add. Regroup if you need to.

1.

	hundreds	tens	ones
	1	1	
	4	6	8
+	3	5	4
	8	2	2

	hundreds	tens	ones
	4	4	7
+		9	6

	hundreds	tens	ones
	6	8	3
+	1	3	5

2.

$$\begin{array}{r} 578 \\ +\,242 \\ \hline \end{array} \qquad \begin{array}{r} 657 \\ +\,173 \\ \hline \end{array} \qquad \begin{array}{r} 436 \\ +\,191 \\ \hline \end{array} \qquad \begin{array}{r} 328 \\ +\,604 \\ \hline \end{array} \qquad \begin{array}{r} 211 \\ +\,499 \\ \hline \end{array}$$

Problem Solving

4. Dryden School recycles containers.

   How many glass bottles and plastic jugs were collected?

   ______ bottles and jugs

Containers Collected	
Item	Number
glass bottles	345
metal cans	624
plastic jugs	120

Spiral Review

Write >, <, or =.

5. 13 _____ 15   59 _____ 47   98 _____ 98   75 _____ 93

Name ____________________

Estimate Sums

Add. Estimate to see if your answer is reasonable.

1.	508 + 291 799	500 + 300 800	187 + 491	+ ___
2.	215 + 279	+ ___	639 + 228	+ ___
3.	423 + 190	+ ___	87 + 540	+ ___
4.	787 + 102	+ ___	536 + 290	+ ___
5.	595 + 341	+ ___	444 + 391	+ ___

Spiral Review

Subtract.

6.

12	15	13	18	14	16
− 9	− 8	− 6	− 10	− 6	− 7

Name ______________________________

Problem Solving: Reading for Math
Main Idea and Details

Henry and Jill paint tiles for a new school mural.

Henry paints 231 tiles. Jill paints 178 tiles.

How many tiles do they paint in all?

1. What is the main idea? ______________________

2. What do you want to find out?

____

3. What details will help you find out?

____

4. How many tiles do they paint in all? ______________

5. Write a number sentence to show your thinking.

____

Spiral Review

Choose + or –.

6. 8 ◯ 6 = 14 16 = 8 ◯ 8 9 ◯ 5 = 4 9 ◯ 6 = 15

Name ____

Problem Solving: Strategy
Make a Graph

Complete the bar graph to solve.

1. In the Jump-Rope-a-Thon, Jim jumped for 10 minutes, hopped for 5 minutes, and skipped for 5 minutes. Mark jumped for 15 minutes and hopped for 10 minutes. Julie skipped for 25 minutes. How many minutes were spent on each of the three activities?

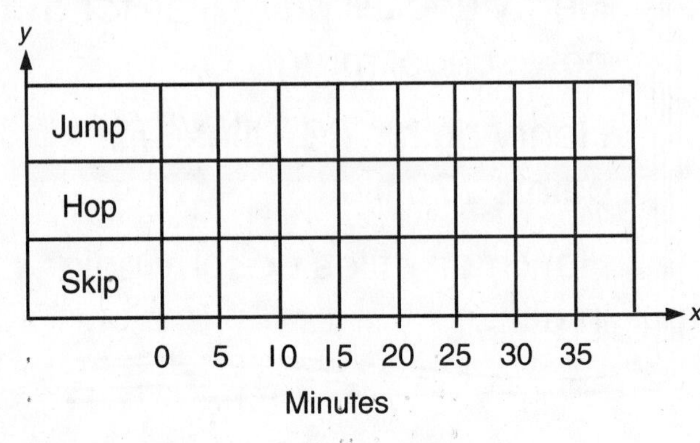

____

2. At the flower sale, Tanya sold 8 red flowers and 2 yellow ones. Dan sold 7 pink flowers and 5 white ones. Marcus sold 3 yellow flowers and 6 blue ones. How many flowers did each of them sell?

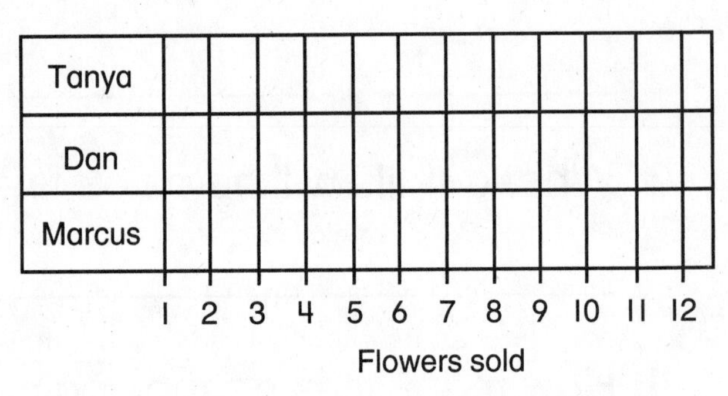

____

Spiral Review

Write the number that comes between.

3. 34 ______ 36 59 ______ 61 100 ______ 102

Name ____________________

Subtract Hundreds

Count back by hundreds to subtract.

1. 600 – 200 = 400 700 – 300 = ____ 300 – 100 = ____

2. 600 – 300 = ____ 900 – 300 = ____ 500 – 400 = ____

3. 800 – 500 = ____ 700 – 400 = ____ 400 – 200 = ____

4.

700	600	300	900	800	600
– 200	– 400	– 300	– 400	– 200	– 500

Problem Solving

5. Find the next number in each pattern.

600 500 400 300 ____ 345 445 545 645 ____

200 300 400 500 ____ 762 662 562 462 ____

6. The third grade washes 200 windows. The second grade washes 400 windows. How many more windows does the second grade wash? ________ windows

Spiral Review

Write each number. Write each word.

8.

____

hundreds	tens	ones
4	5	3

____

____

hundreds	tens	ones
1	6	2

____

Name ___________________________________

3-Digit Subtraction Without Regrouping

Subtract.

1.

	hundreds	tens	ones
	5	5	5
−	4	5	4
	1	0	1

	hundreds	tens	ones
	2	7	6
−		2	5

	hundreds	tens	ones
	2	8	7
−	1	0	6

2. $538 - 201$ $614 - 113$ $936 - 21$ $318 - 200$ $555 - 333$

3. $423 - 10$ $256 - 132$ $767 - 52$ $486 - 113$ $325 - 100$

Solve.

4. What is the same about the numbers 625 and 265? What is different?

____

____

____

____

5. Cathy decorates the lunchroom with 124 red balloons. Bill puts up 257 blue balloons. How many more blue balloons are there than red balloons?

______ balloons

Spiral Review

Write the amount. Use a dollar sign and decimal point.

6.     

$______

Name ____

3-Digit Subtraction

Subtract. Use hundreds, tens, and ones models.

1.

	hundreds	tens	ones
	5	12	
	~~6~~	~~2~~	4
−	3	5	4
	2	7	0

	hundreds	tens	ones
	4	1	6
−	1	2	5

	hundreds	tens	ones
	3	9	5
−	1	0	6

2.

	hundreds	tens	ones
	2	7	1
−	1	6	4

	hundreds	tens	ones
	8	8	2
−	5	3	6

	hundreds	tens	ones
	4	0	7
−	1	3	6

Problem Solving

3. How is 627 − 511 different from 627 − 531?

____

4. Natalie tutors students for 120 minutes on Tuesdays.
She tutors for 480 minutes on Thursdays.
How many more minutes does she tutor on
Thursdays than on Tuesdays? ________ minutes

Spiral Review

Circle each figure that shows equal parts.

Write how many equal parts.

5. 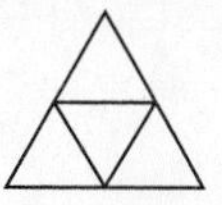 ______ 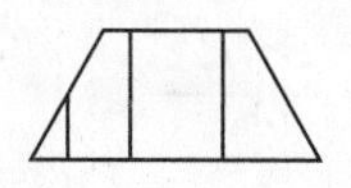 ______ 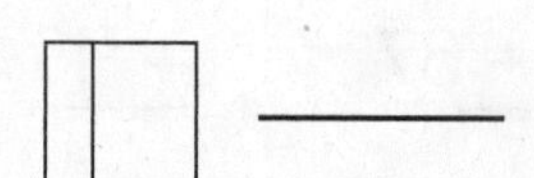 ______

Name ______________________________

More 3-Digit Subtraction

Subtract to find the difference.

1.

	hundreds	tens	ones
	6	11	15
	~~7~~	~~2~~	~~5~~
−	3	2	9
	3	9	6

	hundreds	tens	ones
	☐	☐	☐
	5	2	1
−	1	4	6

	hundreds	tens	ones
	☐	☐	☐
	6	3	2
−	1	5	1

2. $538 - 241$ $\quad 610 - 103$ $\quad 926 - 328$ $\quad 328 - 239$ $\quad 552 - 383$

3. $387 - 226$ $\quad 251 - 182$ $\quad 737 - 259$ $\quad 486 - 193$ $\quad 320 - 145$

Problem Solving

4. Find each missing number.

$572 - \square = 118$ $\quad \square - 225 = 139$ $\quad \square - 360 = 611$

5. There were 345 glass bottles collected in all. 126 of the bottles were green. How many bottles were not green?

______ bottles

Spiral Review

Write the sum.

6. $21 + 37$ $\quad 43 + 49$ $\quad 28 + 38$ $\quad 47 + 22$ $\quad 62 + 8$ $\quad 32 + 45$

Name ____________________

13-12 Estimate Differences

Subtract. Then find the nearest hundred. Estimate each difference to check.

1. 598 − 293 = 305   600 − 300 = 300   687 − 491 = ____   ____ − ____

2. 295 − 279 = ____   ____ − ____   639 − 228 = ____   ____ − ____

3. 423 − 190 = ____   ____ − ____   887 − 540 = ____   ____ − ____

4. 787 − 102 = ____   ____ − ____   536 − 291 = ____   ____ − ____

Problem Solving

5. Tim collects 778 plastic bags. Jack collects 529 paper bags. How many more bags does Tim collect than Jack?

   ______ bags

6. There are 898 flowers. 679 were used to make flower arrangements. How many flowers were not used?

   ______ flowers

Spiral Review

Compare. Use >, <, or =.

7. 27¢ ____ 29¢    ____ 10¢   35¢ ____  

Name ______________________________

13-13 Add and Subtract Money Amounts

Solve. Show your work.

Item	Price
★ Pizza	$1.75
★ Sandwich	$3.55
★ Tacos	$2.25
★ Salad	$2.35
★ Milk	$0.50
★ Juice	$0.85
★ Ice Cream	$1.10

1. Nancy buys pizza and milk. How much does she spend?

$ 2.25

$$\begin{array}{r} \$1.75 \\ +\ 0.50 \\ \hline \$2.25 \end{array}$$

2. Rosa buys a salad. She has $3.00. How much change will she get?

$ ________

3. Marcus has $4.00. Does he have enough to buy tacos and a juice?

____

4. Paige buys a sandwich and milk. How much does her lunch cost?

$ ________

Spiral Review

Count on or back by hundreds. Write each number.

5. One hundred more than 541 is ________.

One hundred less than 732 is ________.

One hundred more than 263 is ________.

Name____

Skip Counting

Skip count by 2s, 5s, or 10s.

Write the numbers.

1.

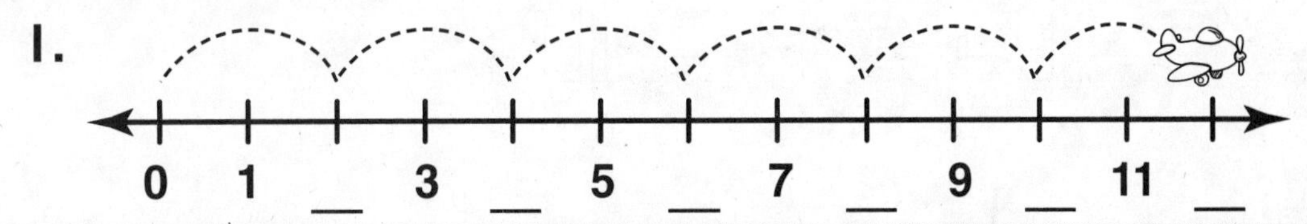

2.

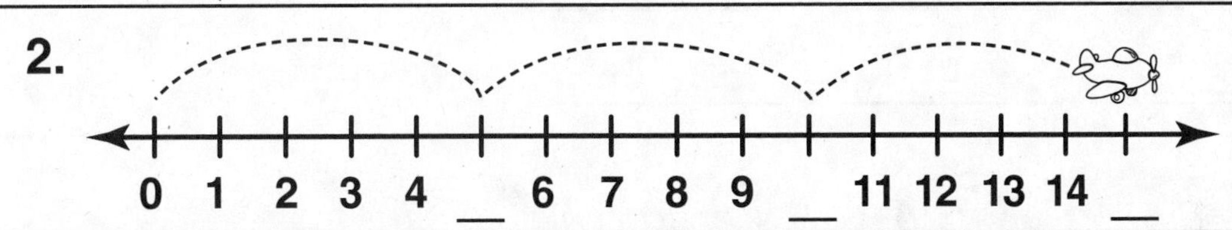

3.

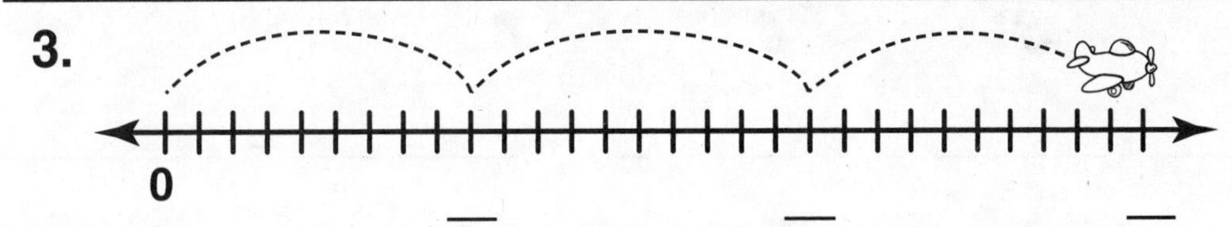

Solve.

4. Look at the picture. What could the next number be? What could the pattern be? Explain.

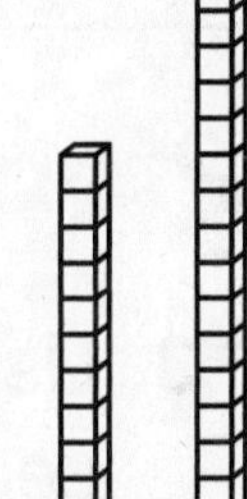

____

____

Spiral Review

Add.

5.

423	256	333	481	325
+ 60	+ 132	+ 52	+ 113	+ 100

Name____

14•2 Repeated Addition and Multiplication

Add. Then multiply.

1. 

4 + 4 = 8

2 × 4 = 8

2.    

_____ + _____ + _____ + _____ = _____

_____ × _____ = _____

Problem Solving

3. Use the addition sentence to complete the multiplication sentence.

4 + 4 + 4 + 4 + 4 = 20

☐ × 4 = 20 What is the missing number? __________

Spiral Review

Add. Estimate to see if your answer is reasonable.

4.

476	587	384
+ 283 + ____	+ 291 + ____	+ 211 + ____

Name ____________________

Use Arrays

Write a multiplication sentence to show each array.

1. $2 \times 3 = 6$

2. ____________

3.  ____________

Find each product.

4. $10 \times 4 =$ ______ $5 \times 7 =$ ______ $2 \times 5 =$ ______

Problem Solving

5. Is 4×5 the same as 5×4?

   ____

   Explain. Then draw a picture to show why or why not.

   ____

Workspace

Spiral Review

Subtract.

6. $71 - 19$ $50 - 35$ $88 - 22$ $67 - 7$ $43 - 16$

Name ______________________

Multiplication

Find each product.

1.

3 × 5 = 15
rows, in each row, in all

____ × ____ = ____
rows, in each row, in all

2.

____ × ____ = ____
rows, in each row, in all

____ × ____ = ____
rows, in each row, in all

3. 2 × 5 = ______ 10 × 3 = ______ 5 × 3 = ______

4. 3 × 2 = ____ 2 × 10 = ____ 6 × 2 = ____ 4 × 5 = ____ 3 × 5 = ____

Problem Solving

5. The marching band has 5 rows.
In each row there are 10 musicians.
How many musicians are there in all? ______ musicians

Spiral Review

Write the time 2 ways.

6. ______ minutes after ______, __________

Name___________________________

More Multiplication

Find each product.

1. 5 groups of 3

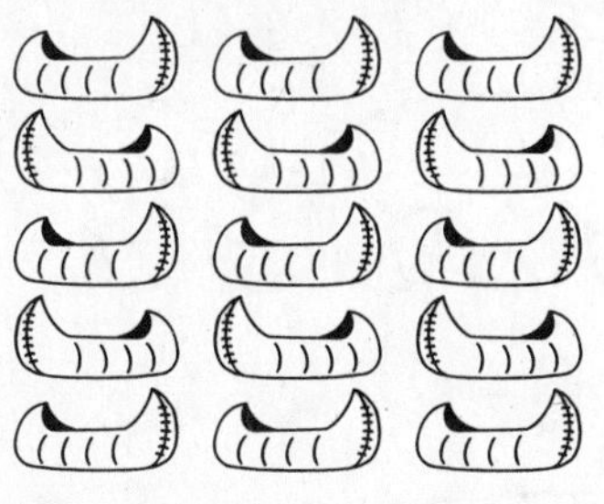

$5 \times 3 =$ ______

3 groups of 5

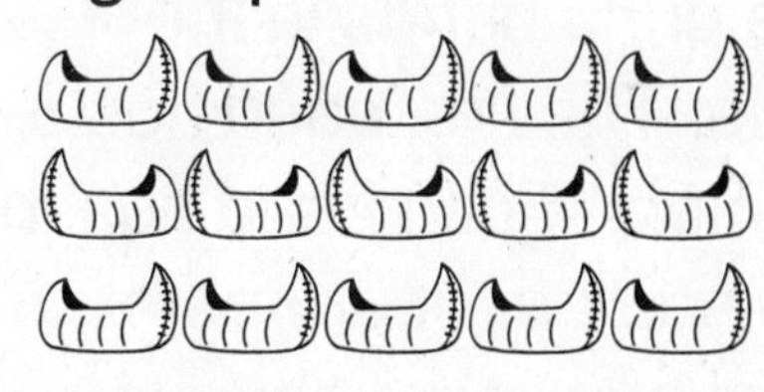

$3 \times 5 =$ ______

2. 2 groups of 3

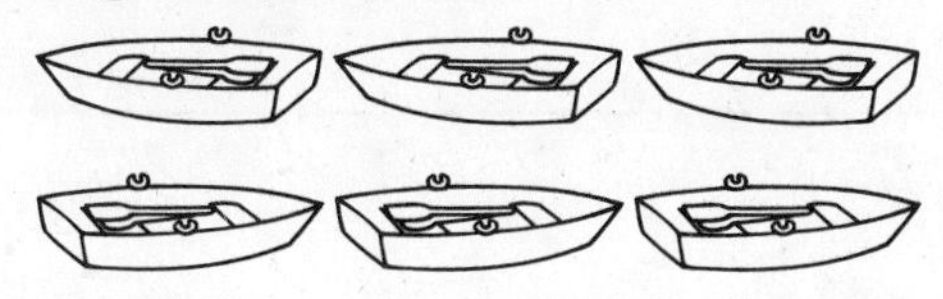

$2 \times 3 =$ ______

3 groups of 2

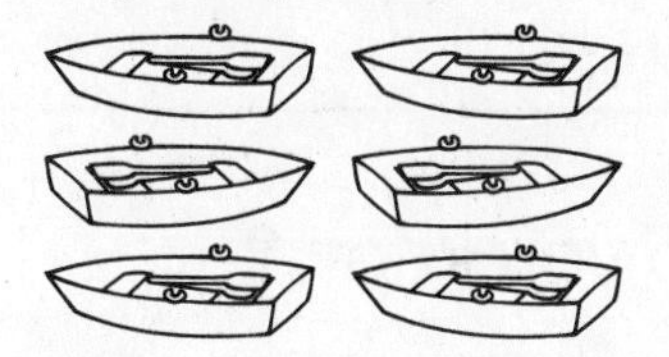

$3 \times 2 =$ ______

3. $2 \times 5 =$ ___ $\quad 5 \times 2 =$ ___ $\quad 6 \times 2 =$ ___ $\quad 2 \times 6 =$ ___ $\quad 8 \times 2 =$ ___ $\quad 2 \times 8 =$ ___

Solve.

There are 9 bicycles at the bike rack.
Each bicycle has 2 wheels. How many wheels in all? ______

Spiral Review

Subtract.

5. $465 - 341 =$ ___ $\quad 788 - 564 =$ ___ $\quad 349 - 214 =$ ___ $\quad 576 - 143 =$ ___ $\quad 639 - 528 =$ ___ $\quad 974 - 241 =$ ___

Name ______________________________

Problem Solving: Reading for Math

Make a Prediction

Vanessa sees a parade. There are 5 bands in the parade. The first band has 6 drummers. Each band has the same number of drummers. Predict how many drummers are in the parade.

1. What do you want to find out?

____

2. What do you know?

____

____

3. How many bands are in the parade? __________

4. Predict the number of drummers in each band. __________

5. How can you predict how many drummers in all? __________

____ bands × ____ drummers in each band = ____ drummers.

Spiral Review

Choose + or –.

6. 24 ◯ 6 = 18   16 = 9 ◯ 7   9 ◯ 5 = 14   9 ◯ 6 = 15

Name ____________________

Problem Solving: Strategy

Draw a Picture

1. 6 buckets of apples are loaded onto each truck. There are 3 trucks. How many buckets are there in all?

18 buckets

2. 4 mail trucks collect the mail. Each truck collects 5 bags of mail. How many bags of mail are there in all?

______ bags

3. There are 7 milk trucks. Each truck delivers 10 crates of milk every morning. How many crates of milk are delivered every morning?

______ crates

Spiral Review

Count back by 4s.

5. 33, 29, ______, ______, ______, ______, ______

Name ______________________________

Repeated Subtraction

How many equal groups can you make?

1. Use 12 [cube].
Subtract groups of 4.

You get ___3___ groups of 4.

Use 18 [cube].
Subtract groups of 6.

You get ______ groups of 6.

2. Use 9 [cube].
Subtract groups of 3.

You get ______ groups of 3.

Use 20 [cube].
Subtract groups of 5.

You get ______ groups of 5.

3. Use 15 [cube].
Subtract groups of 3.

You get ______ groups of 3.

Use 16 [cube].
Subtract groups of 4.

You get ______ groups of 4.

Problem Solving

4. Mrs. Arnold's class is taking a canoe trip. They need to carry 18 canoes to the river. Each trailer holds 6 canoes. How many trailers are needed to carry the canoes? ________ trailers

Spiral Review

Add.

5.

423	256	333	481	325
+ 60	+ 132	+ 52	+ 113	+ 100

Name ____________________

Subtraction and Division

Subtract. Then divide.

1. There are 24 people who want to go sailing.
   Each sailboat can carry 6 people.

   How many times can you subtract 6? ___4___

   24 ÷ 6 = ________

   How many sailboats are needed? ________

Draw a picture and solve.
Use repeated subtraction. Divide.

2. 25 people
   5 in each car

   25 ÷ 5 = ________

   How many cars are needed?

   _____ cars

Workspace

3. 21 goats
   3 in each truck

   21 ÷ 3 = ________

   How many trucks are needed?

   ________ trucks

Workspace

Spiral Review

Write the number that comes just before.

4. _____ 397 _____ 288 _____ 343 _____ 659

Name ______________________________

14·10 Division

Make equal groups. Divide.
Write how many in each group.

1. 9 sailboats — $9 \div 3 =$ 3

   3 groups — 3 in each group

---

2. 8 tow trucks — $8 \div 4 =$ ______ in each group

   4 groups — ______ in each group

Circle *Fair* or *Not Fair*.

3. There are 8 oranges for 4 friends.
   Joel takes 3 oranges. Fair Not Fair

4. There are 5 baskets for 30 apples.
   One basket has 6 apples. Fair Not Fair

Problem Solving

5. A farmer has 10 pigs. There are 3 trucks to carry them. Can each truck carry the same number of pigs? ______

   Explain: ______________________________

Spiral Review

Add. Regroup if you need to.

6.

608	783	568	618	289	189
+ 139	+ 143	+ 376	+ 109	+ 418	+ 463

Name ______________________________

More Division

Make equal groups. Write how many are in each group and how many are left over.

1. 26 baseballs
3 buckets

__8__ baseballs in each bucket

__2__ baseballs left over

Workspace

2. 26 fish
4 tanks

______ fish in each tank

______ fish left over

Workspace

Divide into equal groups. Write how many in each group. Write how many are left over.

3. 14 into 7 equal groups ______ in each group ______ left over

4. 9 into 2 equal groups ______ in each group ______ left over

Solve.

5. A farmer has 32 oranges and 4 crates to pack them in. How many can go into each crate if she makes equal groups? _____

Spiral Review

Write The number that comes in between.

6. 659 ______ 661 599 ______ 601 428 ______ 430